Andrew Browne

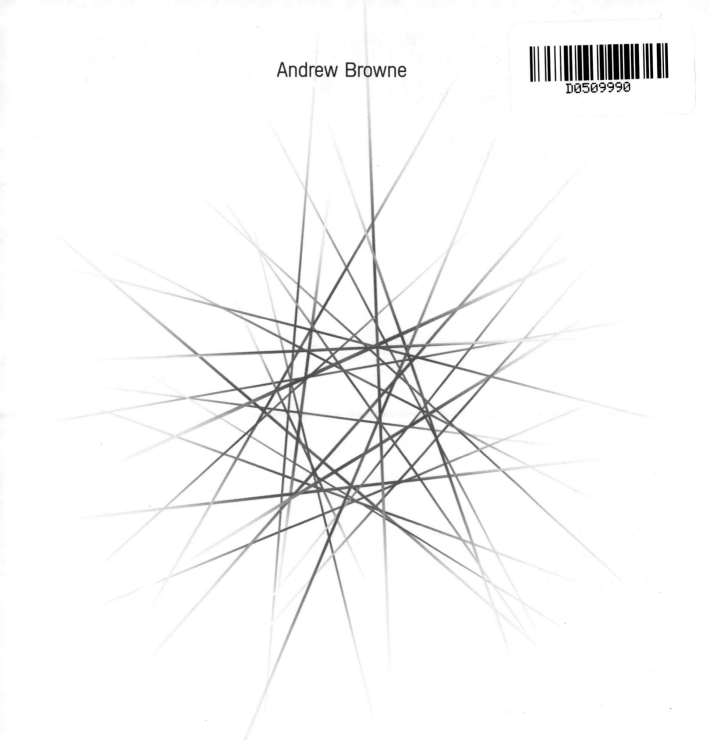

ESSENTIALS

AQA Specification A
GCSE Geography
Workbook

Contents: Physical Geography

Contents: Human Geography

Population and Urban Environments

Rural Environments and Development

Globalisation and Tourism

The Earth's Crust

Plates

1 Circle the correct options in the following sentences.

a) The Earth's crust is broken into **7 / 12** major plates.

b) Oceanic plates are made of **basalt / granite**.

c) Continental plates are **5−10km / 25−90km** thick.

d) Oceanic plates are **denser / lighter** than continental plates.

2 The diagram shows the movement of the Earth's plates. Match statements **A**, **B**, **C** and **D** with the labels **1−4** on the diagram. Enter the appropriate number in the boxes provided.

A Mantle

B Crust

C Core

D Convection currents

Fold Mountains

3 Choose the correct words from the options given to complete the following sentences.

anticlines	**synclines**	**destructive**	**constructive**

Fold mountains are formed when two plates meet at a _____ plate margin. Where the

land is folded upwards, _____ are formed. Where the land has been folded

downwards, the folds are known as _____.

4 What is a geosyncline?

_____

_____

_____

The Earth's Crust

Destructive Plate Margins

5 Which of the following statements about destructive plate margins are true? Tick the correct options.

A Oceanic and continental plates are moving apart. ◯

B Oceanic crust is forced down into the mantle. ◯

C Continental crust is compressed to form fold mountains. ◯

D There are no volcanoes. ◯

Constructive Plate Margins

6 Fill in the missing words to complete the following sentences.

Constructive plate margins are where continental or oceanic plates are pulling ....

Where lava erupts on land, a .... volcano is formed. Where lava erupts under the

sea, a .... is formed.

Conservative Plate Margins

7 What is a fault line?

....

....

....

8 Why do major earthquakes occur at conservative plate margins?

....

....

....

....

Volcanoes

Types of Volcano

1 What is a hotspot?

_____

2 Which of the following statements about basic lava are true? Tick the correct options.

A It is 1200°C. ◯

B It is 800°C. ◯

C It is thick and sticky. ◯

D It is thin and runny. ◯

3 Circle the correct option in the following sentence.

An example of a shield volcano is **Mauna Loa / Mt St Helens**.

4 Fill in the missing words to complete the following sentences.

A composite volcano is found on a _____ plate margin. The type of lava that erupts

from a composite volcano is known as _____ lava. Composite volcanoes are

cone-shaped with _____ sloping sides. An example of a composite volcano

is _____.

5 Which of the following is a sunken crater formed after a volcanic eruption? Tick the correct option.

A Cirque ◯

B Caldera ◯

C Cravasse ◯

D Cavern ◯

6 a) When was the most recent supervolcano eruption? _____

b) Where was the eruption? _____

c) How would a supervolcano eruption affect people and the planet today? _____

_____

_____

Monitoring Volcanoes

7 The table contains four terms related to monitoring volcanoes.

Match descriptions **A**, **B**, **C** and **D** with the terms **1–4** in the table. Enter the appropriate number in the boxes provided.

1	Seismometers
2	Remote sensing
3	Tiltmeters
4	Volcanologists

A Scientists who specialise in studying volcanoes.

B The use of satellites to monitor volcanoes.

C Equipment used to measure how the shape of volcanoes changes over time.

D Equipment that measures the earthquake activity that occurs before an eruption.

Case Study: Montserrat

8 Choose the correct words from the options given to complete the following sentences.

ash   **destructive**   **lava**   **pyroclastic flows**   **composite**   **constructive**

The island of Montserrat is on a _____ plate boundary, where the South American

and Caribbean plates meet. Between 1995 and 1997, two thirds of the island were covered by

_____ and 60% of the houses were destroyed by _____

_____.

The eruption was particularly violent because the Soufriere Hills is a _____ volcano.

9 Describe two primary effects of the Montserrat volcano eruption.

a) _____

b) _____

Earthquakes

The Cause of Earthquakes

1 The table contains four terms related to earthquakes. Match descriptions **A, B, C** and **D** with the terms **1–4** in the table. Enter the appropriate number in the boxes provided.

1	Focus
2	Epicentre
3	Shock waves
4	Richter scale

A   The point on the ground surface directly above an earthquake.

B   Measures the amount of energy released by an earthquake.

C   The point underground where an earthquake begins.

D   Energy released by an earthquake that travels outwards from the epicentre.

2 Circle the correct options in the following sentences.

a)   The energy released by an earthquake is plotted on a **seismometer / seismograph**.

b)   The scale used to measure the amount of damage caused by an earthquake is known as the **Mercalli / Richter** scale.

Case Study: Pakistan, 2005

3 Where was the epicentre of the 2005 Pakistan earthquake? ....

4 Choose the correct numbers from the options given to complete the following sentences.

3 million	73 000	1 million	7.6

The energy released by the earthquake was measured at .... on the Richter scale.

The number of people who died was .... , and .... were made homeless.

Case Study: Kobe, Japan, 1995

5 Which of the following statements about the 1995 Japanese earthquake are true? Tick the correct options.

A   The earthquake measured 7.2 on the Richter scale.

B   The epicentre was on Awaji Island.

C   The earthquake was caused by a conservative plate boundary.

Case Study: Asian Tsunami, 2004

6 Which of the following was the Richter scale measurement for the Asian Tsunami? Tick the correct option.

A 5.6 ⃝

B 7.2 ⃝

C 7.6 ⃝

D 9.2 ⃝

7 a) How long was the break in the crust that caused the earthquake? ....

b) How high was the sea bed raised by the earthquake? ....

c) How fast did the waves travel across the Indian Ocean? ....

d) How high did the waves rise? ....

8 Fill in the missing words to complete the following sentences.

The tsunami killed approximately 230 000 people and left .... homeless.

Over £ .... was donated by governments and people. The United Nations has

coordinated the development of a tsunami .... for the Indian Ocean.

Prediction and Preparation

9 What is available to help people prepare for earthquakes?

a) ....

b) ....

c) ....

⃝

Rocks and Weathering

Geological Timescale

1 The diagram shows the geological timescale. Match the descriptions **A**, **B**, **C** and **D** with the labels **1–4** on the diagram. Enter the appropriate number in the boxes provided.

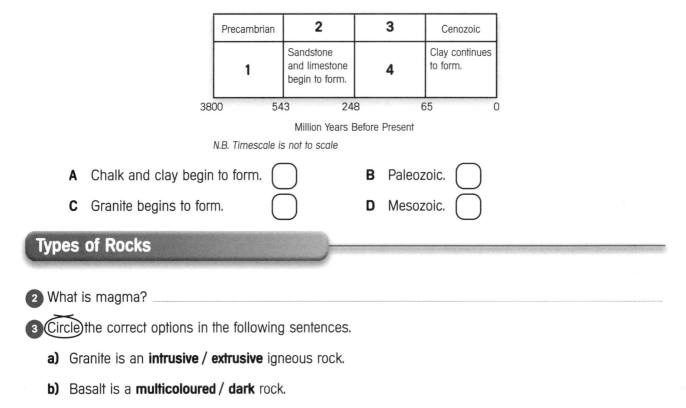

Precambrian	**2**	**3**	Cenozoic
1	Sandstone and limestone begin to form.	**4**	Clay continues to form.

3800   543   248   65   0

Million Years Before Present

N.B. Timescale is not to scale

A Chalk and clay begin to form. ◯

B Paleozoic. ◯

C Granite begins to form. ◯

D Mesozoic. ◯

Types of Rocks

2 What is magma? ....

3 Circle the correct options in the following sentences.

a) Granite is an **intrusive / extrusive** igneous rock.

b) Basalt is a **multicoloured / dark** rock.

c) Granite has a **coarse / fine** texture.

4 The table contains the names of four sedimentary rocks. Match descriptions **A**, **B**, **C** and **D** with the terms **1–4** in the table. Enter the appropriate number in the boxes provided.

1	Chalk
2	Limestone
3	Sandstone
4	Clay

A Grains of sand cemented together. ◯

B Shells of tiny sea creatures. ◯

C Particles of silt and clay. ◯

D Coral and shells of sea creatures. ◯

5 How are metamorphic rocks formed?

....

....

Rocks and Weathering

Types of Rocks (cont.)

6 Draw lines between the boxes to match each type of metamorphic rock to the sedimentary rock it's made from.

Marble		Sandstone
Slate		Chalk and limestone
Quartzite		Clay and mudstone

Biological Weathering

7 Describe how tree and plant roots are able to break down rocks.

Chemical Weathering

8 Acid rain is able to dissolve rocks that contain calcium carbonate. Which of the following rocks is made from calcium carbonate? Tick the correct option.

A Granite

B Limestone

C Basalt

D Quartzite

Mechanical Weathering

9 Fill in the missing words to complete the following sentences about freeze-thaw weathering.

a) Water seeps into .... in rocks.

b) Water freezes, turns into ice and ...., which breaks down the rock.

c) The ice .... and the process begins again.

10 Explain how heating and cooling can cause the breakdown of rocks.

Landscapes, Landforms and Resources

Granite Landscapes

1 Which of the following is the name for an intrusion of igneous rock underground? Tick the correct option.

A Megalith ◯   **B** Monolith ◯

C Hieroglyph ◯   **D** Batholith ◯

2 Which of the following statements about granite landscapes are true? Tick the correct options.

A Moorland is formed from granite, weathered over thousands of years. ◯

B Weathering reveals joints and blocks in granite called grikes and clints. ◯

C Granite is impermeable, which means it doesn't let water pass through it. ◯

D Tors are blocks of granite. ◯

E Granite is dissolved by rainwater. ◯

F Granite landscapes contain many rivers and marshes. ◯

3 Write down one advantage and one disadvantage of granite landscapes in terms of opportunities for people.

a) Advantage: ....

b) Disadvantage: ....

Chalk and Clay Landscapes

4 The table contains four terms relating to chalk and clay landscapes.

Match descriptions **A, B, C** and **D** with the terms **1–4** in the table. Enter the appropriate number in the boxes provided.

1	Vale
2	Escarpment (Cuesta)
3	Dry river valley
4	Scarp slope

A A steep sloping hillside that marks the boundary between chalk and clay. ◯

B A valley cut into an escarpment during the last ice age. ◯

C Gently sloping hills formed from more resistant chalk. ◯

D A flat plain formed from less resistant clay. ◯

Landscapes, Landforms and Resources

Carboniferous Limestone

5 The diagram shows a limestone landscape. Match the statements **A, B, C, D** and **E** with the labels **1−5** on the diagram. Enter the appropriate number in the boxes provided.

A Limestone pavement ☐   **B** Swallow hole ☐

C Cavern ☐   **D** Stalactites ☐

E Stalagmites ☐

Case Study: Yorkshire Dales

6 Describe two ways in which limestone quarrying can damage the environment.

a) ....

b) ....

7 Fill in the missing words to complete the following sentences.

a) Quarrying provides .... for local people and raw .... for industry.

b) After quarrying, the landscape is restored; at Swinden, the limestone quarry will be replaced

with a .... and a ....

Weather and Climate

UK Weather and Climate

1 Which of the following describes the UK's climate? Tick the correct option.

A Temperate ◯   **B** Sub-tropical ◯

C Tropical ◯   **D** Arid ◯

2 Circle the correct option in the following sentence.

The UK's climate is affected by the sea that surrounds the country. Therefore, the UK is described as having a **maritime / continental** climate.

Temperatures and Precipitation

3 Circle the correct options in the following sentences.

a) In **July / January**, the main influence on the UK's temperature is latitude.

b) The wettest parts of the UK are on the **east / west** coast.

Factors Affecting Climate

4 The table contains five terms relating to factors that affect climate. Match descriptions **A, B, C, D** and **E** with the terms **1–5** in the table. Enter the appropriate number in the boxes provided.

1	Latitude
2	Altitude
3	Pressure
4	Wind
5	Continentality

A The temperature and moisture content of air blowing towards a country. ◯

B The distance from the equator. ◯

C The distance from the sea. ◯

D The height of land above sea level. ◯

E The movement of air rising or falling above a country. ◯

Weather and Climate

Depressions

5 Choose the correct words from the options given to complete the following sentences.

| **warm** | **cold** | **pressure** | **front** | **precipitation** |

a) Depressions are areas of low _____ that form when warm air rises over cold air.

b) Where warm air meets cold air, a _____ is formed.

c) Warm air rising over cold air is known as a _____ front. Cold air undercutting

warm air from behind is known as a _____ front.

d) As the warm air rises over the cold air, it cools, condenses and _____ occurs.

6 Using the diagram to help you, describe how the weather changes as a depression moves overhead.

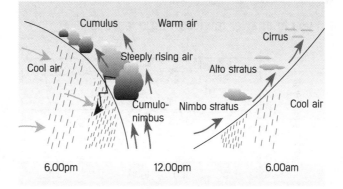

a) 6.00am: _____

b) 12.00pm: _____

c) 6.00pm: _____

Anticyclones

7 Which of the following statements about anticyclones are true? Tick the correct options.

A Anticyclones are areas of high pressure.

B Anticyclones are formed when air is sinking.

C In the summer, anticyclones bring cloud and rain.

D In the winter, anticyclones bring sunny and frosty weather.

E Winds blow in an anticlockwise direction.

Global Climate Change

The Greenhouse Effect

1 The table contains three terms relating to the greenhouse effect. Match descriptions **A**, **B** and **C** with the terms **1–3** in the table. Enter the appropriate number in the boxes provided.

1	Greenhouse effect
2	Shortwave radiation
3	Longwave radiation

A Solar radiation that passes through the atmosphere and warms the Earth's surface.

B The process of gases in the Earth's atmosphere absorbing heat radiated from Earth.

C Heat radiated from the Earth.

Evidence for Global Climate Change

2 Use the diagram to help you to answer the following questions.

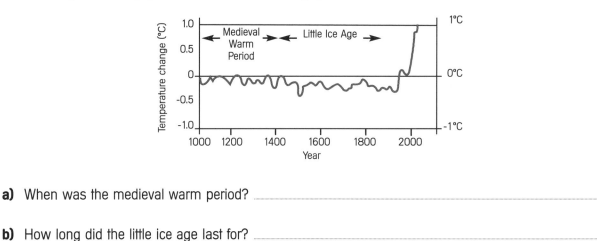

a) When was the medieval warm period? ....

b) How long did the little ice age last for? ....

c) By how much has the global temperature risen since 1900? ....

3 Which of the following can be used to measure changes in global atmospheric carbon dioxide and temperature over thousands of years? Tick the correct option.

    A Tree rings ◯   **B** Continental drift ◯   **C** Ice cores ◯

Causes of Global Climate Change

4 Fill in the missing words to complete the following sentences about the causes of climate change.

Carbon dioxide is emitted by burning fossil fuels. .... is released by

millions of cattle every day. .... cause an increase in solar activity.

Global Climate Change

Consequences of Global Climate Change Revision Guide Reference: Page 19

5 a) How many people could be made homeless if sea levels rise by 1 metre?

....

b) Which two categories of hurricane are becoming more frequent?

i) .... **ii)** ....

c) Name a tropical disease that could become more common in Europe.

....

d) How many people could face water shortages in Africa by 2020?

....

e) If temperatures rise by 1.5°C, what percentage of plant and animal species could become extinct?

....

Responses to Global Climate Change

6 Fill in the missing words to complete the following sentences.

a) An agreement made in 1997 by 182 countries to reduce carbon emissions is known as the

....

b) Countries that emit fewer greenhouse gases than others are able to sell ....

.... to countries that want to emit more greenhouse gases than they're permitted to.

c) Sources of energy, such as wind power and hydroelectricity, are known as .... energy.

d) Some cities have introduced .... to encourage people to use public transport.

e) Loft insulation, cavity wall filling and low-energy light bulbs are all examples of

....

Extreme Weather

Tropical Storms

1 Which of the following statements about tropical storms are true? Tick the correct options.

A They form in late summer.

B They're known as hurricanes, cyclones, typhoons and willy willies.

C They move in an easterly direction.

D They are high pressure systems.

E Winds up to 200km/h can destroy property and can cause flooding and landslides.

F A storm surge is where tropical storms can cause sea levels to rise by several metres.

2 The diagram shows a tropical storm. Match the statements **A, B, C, D** and **E** with the labels **1−5** on the diagram.

A Eye

B Canopy

C Strong winds

D Gentle winds

E Direction of rotation

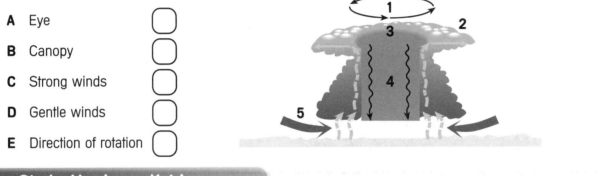

Case Study: Hurricane Katrina

3 The table contains statistics relating to Hurricane Katrina. Match descriptions **A, B, C** and **D** with the statistics **1−4** in the table. Enter the appropriate number in the boxes provided.

1	6
2	80
3	1836
4	80 billion

A The cost of the damage (in US dollars).

B The height of the storm surge (in metres).

C The number of people killed by Hurricane Katrina.

D The percentage of New Orleans that was flooded.

4 Why were the state and the national government heavily criticised for their response to the disaster?

Case Study: Cyclone Nargis

5 (Circle) the correct options in the following sentences.

a) **65% / 95%** of buildings in the Irrawaddy delta were destroyed.

b) **13 300 / 133 000** people were reported dead or missing.

c) Over **100 000 / 1 million** people were made homeless.

d) The cost of the cyclone was **$10bn / $100bn**.

6 Why did people in Burma receive so little aid?

....

....

....

Extreme Weather in the UK

7 Which of the following was the warmest spring on record in the UK? Tick the correct option.

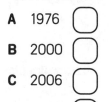

A 1976 ⬡

B 2000 ⬡

C 2006 ⬡

D 2007 ⬡

8 The summer of 2007 was one of the wettest on record in the UK.

a) How many buildings were flooded? ....

b) How many people had their water supply cut off? ....

c) What was the cost of the damage caused by the flooding? ....

Ecosystems

Ecosystems

1 What is an ecosystem?

....

....

Nutrient Cycles

2 The diagram shows a nutrient cycle. Match the statements **A, B, C** and **D** with the labels **1–4** on the diagram. Enter the appropriate number in the boxes provided.

    A Decomposition of plants and animals into humus. ◯

    B Weathering releases nutrients from rocks. ◯

    C Plants absorb nutrients through their roots. ◯

    D Leaves and branches fall. ◯

Energy Flows

3 Choose the correct words from the options given to complete the following sentences.

    producers   **consumers**   **chain**   **photosynthesis**

    a) Energy flows through an ecosystem along a food ....

    b) Plants convert solar energy using a process called ....

    c) Plants are known as ...., because they're able to convert the sun's energy.

    d) Animals are known as ...., because they receive energy by eating plants and other animals.

Human Modificiation

4 Fill in the missing words to complete the following sentences.

    a) Introducing a new .... of plant or animal can upset the balance of an ecosystem.

    b) Removing vegetation or adding nutrients to an ecosystem during .... will modify the way an ecosystem functions.

Case Study: Decidous Woodland

5 Which of the following continents does not contain deciduous woodland? Tick the correct option.

A Europe ⬭

B Asia ⬭

C Africa ⬭

D North America ⬭

6 Circle the correct options in the following sentences about temperate deciduous woodland.

a) It's found in locations with **mild / cold** winters and **warm / hot** summers.

b) Precipitation falls all year round, with an average annual precipitation of **100mm / 1000mm**.

c) Soil type is described as **brown soil / red clay**.

7 a) Name three species of tree that are found in deciduous woodland.

i) ....

ii) ....

iii) ....

b) Name two species of shrub that are found in deciduous woodland.

i) ....

ii) ....

c) Name one other plant that's found in deciduous woodland.

....

Human Activity on Deciduous Woodland

8 Which of the following describes the practice of coppicing? Tick the correct option.

A Cutting trees down to ground level to encourage shoots to regrow, which are then harvested. ⬭

B Felling trees to provide timber for industry and the construction of furniture. ⬭

C A system of managing woodland to provide habitats for wildlife, and recreation for people. ⬭

Managing Ecosystems

Case Study: Tropical Rainforests

1 Describe the location of the world's tropical rainforests.

....

2 Which of the following is a description of the climate of a tropical rainforest? Tick the correct option.

    A Mild temperatures (4°C–6°C); precipitation all year (1000mm per year).   ◯

    B High temperatures all year (30°C); very dry (250mm per year).   ◯

    C High temperatures all year (27°C); precipitation daily (2000mm per year).   ◯

3 Fill in the missing words to complete the following sentences.

  a) Trees have wide .... roots to support their 30–50m height.

  b) Leaves have a .... surface to help water to drain off.

  c) Roots are .... to absorb nutrients from decomposing vegetation.

Causes and Impacts of Deforestation

4 Describe one cause of deforestation.

....

5 The table contains four terms relating to the impacts of deforestation. Match descriptions **A**, **B**, **C** and **D** with the terms **1–4** in the table. Enter the appropriate number in the boxes provided.

1	Economic
2	Social
3	Political
4	Environmental

    A Soil erosion, species loss and climate change.   ◯

    B Short-term gain of revenue from selling timber.   ◯

    C Governments want their countries to develop, but want to protect rainforests at the same time.   ◯

    D Indigenous rainforest tribes are losing their homes and culture.   ◯

Managing Ecosystems

Sustainable Management

6 Which of the following are sustainable methods of managing tropical rainforests? Tick the correct options.

A Selective felling of trees followed by replanting ⬭

B Cattle ranching ⬭

C Encouraging ecotourism ⬭

D Open-cast bauxite mining ⬭

E Reducing demand for tropical hardwoods in richer countries ⬭

Case Study: Hot Deserts

7 Circle the correct options in the following sentences.

a) Hot deserts are found mainly on the **tropics of Cancer and Capricorn / equator**.

b) Hot deserts have less than 250mm precipitation a year due to **low / high** pressure.

c) Plants that can survive in an environment with very little water are known as **xerophytic / halophytic**.

d) Plants that can store water in their stems and leaves are known as **ungulates / succulents**.

Human Activity in Hot Deserts

8 Give one reason why so many people visit Las Vegas in the Mojave Desert each year.

....

....

9 The traditional culture of the Kalahari Desert bushmen in Botswana is under threat from mining for which mineral? Tick the correct option.

A Gold ⬭

B Copper ⬭

C Diamonds ⬭

D Tin ⬭

River Processes

Discharge and Velocity

1 Circle the correct options in the following sentences.

Discharge is the amount of water in a river; it is measured in **litres per second / cubic metres per second**. Discharge **increases / decreases** from the source to the mouth.

2 The table contains six factors that affect the discharge of a river. Match descriptions **A, B, C, D, E** and **F** with the terms **1–6** in the table. Enter the appropriate number in the boxes provided.

1	Rainfall
2	Temperature
3	Previous weather
4	Relief
5	Rock type
6	Land use

A Impermeable or permeable ⬭   **B** Type and amount ⬭

C Steep or gentle ⬭   **D** Rural or urban ⬭

E Wet or dry ⬭   **F** Hot or cold ⬭

3 Fill in the missing words to complete the following sentences.

Velocity is the speed of a river; it is measured in _____. Velocity usually

_____ from source to mouth. This is because there's more water in the river,

so there's less _____ from the river bed and banks.

Erosion

4 Choose the correct words from the options given to complete the following sentences.

        action   **abrasion**   **attrition**   **corrosion**

a) _____ is the erosion of the river channel by stones being transported by the river.

b) The erosion of the river channel by the force of flowing water is known as hydraulic _____.

c) If the river water is slightly acidic, rocks that are made from calcium carbonate are dissolved

by _____.

Transportation

5 The diagram shows the four ways a river transports its load. Match terms **A, B, C** and **D** with the labels **1–4** on the diagram. Enter the appropriate number in the boxes provided.

A Traction ◯

B Saltation ◯

C Solution ◯

D Suspension ◯

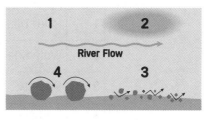

Deposition

6 Which of the following statements about deposition are true? Tick the correct options.

A Material is deposited when rivers slow down. ◯

B Minerals transported in solution become salt in the sea. ◯

C The lightest material gets deposited first, followed by the heaviest. ◯

D Deposition happens on the inside of meanders, where the river has less energy to transport material. ◯

Long Profiles

7 Which of the following describes the long profile of a river? Tick the correct option.

A The gradient increases gradually downstream. ◯

B A steep upper course, with a gentle lower course. ◯

C Convex shape. ◯

Valley Cross-profiles

8 Draw lines between the boxes to match each course of a river with its description.

| Upper | | Vertical and lateral erosion, gently sloping sides. Bedload: pebbles, stones. |

| Middle | | Lateral erosion, wide valley floor. Bedload: sand, silt, clay. |

| Lower | | Vertical erosion, steep valley sides. Bedload: boulders, stones. |

River Landforms

River Basins and Features of Rivers

1 The table contains six features of a river basin. Match descriptions **A**, **B**, **C**, **D**, **E** and **F** with the terms **1–6** in the table. Enter the appropriate number in the boxes provided.

1	River basin
2	Watershed
3	Source
4	Tributary
5	Confluence
6	Mouth

A   Where a river begins.   ◯

B   The point where two rivers join.   ◯

C   Where a river enters a lake or the sea.   ◯

D   High land that surrounds a river basin.   ◯

E   A small river that joins a larger river.   ◯

F   The area drained by a river and its tributaries.   ◯

Waterfalls and Meanders

2 The diagram shows a waterfall and its gorge. Match statements **A**, **B**, **C**, **D**, **E**, **F** and **G** with the labels **1–7** on the diagram. Enter the appropriate number in the boxes provided.

A   Gorge   ◯

B   Plunge pool   ◯

C   Hard rock   ◯

D   Soft rock   ◯

E   Waterfall retreats upstream   ◯

F   Direction of river flow   ◯

G   Waterfall   ◯

3 a)   Where is the fastest flow of water found in a meander? ....

b)   Where does deposition occur in a meander? ....   ◯

River Landforms

Ox-bow Lakes

Fill in the missing words to complete the following sentences.

An ox-bow lake is formed when a river _____ through the neck of a meander during

a period of high flow. The ends of the meander are filled in by _____, leaving a

_____-shaped lake next to the river.

5 Name an example of an ox-bow lake.

_____

Flood Plains

6 Which of the following describes a flood plain? Tick the correct option.

A A place where water flows vertically. ◯

B A curve or loop in a river. ◯

C A flat area of land formed on either side of a river. ◯

D An area drained by a river and its tributaries. ◯

Levees

7 Circle the correct options in the following sentences.

a) Levees are raised banks of sediment that have been **eroded / deposited** on either side of a river channel.

b) Levees are formed from the **largest / smallest** particles of sediment.

c) Levees develop over many years of river **flooding / meandering**.

8 Name an example of a levee.

_____

◯

River Flooding and River Management

Causes of Flooding

1 Which of the following are causes of river flooding? Tick the correct options.

A Planting trees ⬭

B Melting snow ⬭

C Building towns and cities ⬭

D Flat land ⬭

E Heavy rain ⬭

F A prolonged period of rain ⬭

Case Study: Boscastle Floods

2 Choose the correct numbers from the options given to complete the following sentences.

| 100 | 58 | 2004 | 77 | 300 | 2 |

a) The village of Boscastle was flooded in July _____. The ground was waterlogged

from previous rainfall, then storms caused _____ mm of rain to fall in

_____ hours.

b) Steep slopes and narrow valleys channelled the water through the village, causing

£_____ million worth of damage.

c) _____ properties were flooded and _____ people were airlifted
to safety.

Case Study: Mozambique Floods

3 a) In which year did the floods occur? _____

b) How long did it rain for in February? _____

c) What type of extreme weather followed the rainfall? _____

d) How many people were killed in the floods? _____

e) How many people were made homeless by the floods? _____

River Flooding and River Management

Flood Management

4 Circle the correct options in the following sentences.

a) The use of structures or machinery to control flooding is known as **hard / soft** engineering.

b) Rivers can be straightened to **increase / decrease** the flow of water to prevent flooding.

c) Working with the environment is known as **hard / soft** engineering.

d) Preventing house building on flood plains is an example of flood **warning / zoning**.

Water Supply

5 Name the three main sources of water supply in the UK.

a) ....

b) ....

c) ....

6 Which of the following are reasons for an increase in the demand for water in the South East of England? Tick the correct options.

A Fewer farms ⬭

B Increase in house building ⬭

C Decline in leisure and tourism ⬭

D Increase in industrial development ⬭

7 Fill in the missing words to complete the following sentences.

a) Thames Water wants to build a new reservoir in .... The reservoir will cost £1 billion and will flood an area of .... square miles.

b) Thames Water says the reservoir will provide .... with 350 million litres of water a day. Opponents say high quality .... will be flooded and that Thames Water should spend its customers' money on saving the 900 million litres of water lost a day from ....

Glaciation

Ice Ages

1. What is an ice age?

   ....

2. Circle the correct options in the following sentences.

   a) The most recent ice age began **200 000 / 2 million** years ago.

   b) Warmer periods within an ice age are known as **glacials / inter-glacials**.

   c) The last glacial ended **10 000 / 100 000** years ago.

   d) During the last glacial, ice covered about **10% / 30%** of the Earth's surface.

   e) Today, large ice masses are found at **high / low** latitudes.

Glacial Processes and Glacial Budget

3. The table contains five glacial processes. Match descriptions **A, B, C, D** and **E** with the terms **1–5** in the table. Enter the appropriate number in the boxes provided.

1	Freeze-thaw weathering
2	Lodgement
3	Plucking
4	Abrasion
5	Ablation

   A Rocks carried in the ice erode the valley sides and floor. ⬭

   B Loose rocks become frozen into the glacier and are pulled away as the glacier moves downhill. ⬭

   C Deposition when a glacier becomes overloaded with debris. ⬭

   D Deposition when a glacier melts. ⬭

   E Water gets into cracks in rocks, then freezes, expands and shatters the rocks. ⬭

4. Draw lines between the boxes to match each term to its meaning.

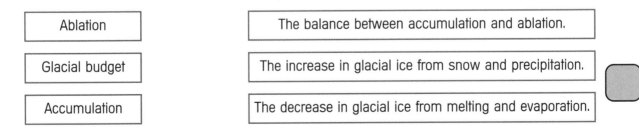

Ablation		The balance between accumulation and ablation.
Glacial budget		The increase in glacial ice from snow and precipitation.
Accumulation		The decrease in glacial ice from melting and evaporation.

Glaciation

Case Study: Athabasca Glacier, Canada

5 The diagram shows a glacier in cross section. Match statements **A, B, C** and **D** with the labels **1–4** on the diagram. Enter the appropriate number in the boxes provided.

A Meltwater ◯

B Equilibrium line ◯

C Ablation zone ◯

D Accumulation zone ◯

Case Study: Chamonix Valley, French Alps

6 Which of the following statements about tourism in the French Alps are true? Tick the correct options.

A People visit the French Alps to view the stunning scenery. ◯

B People are attracted to the French Alps by the Mediterranean climate. ◯

C Skiing is the main reason why tourists visit the French Alps between December and May. ◯

D The French Alps provide opportunities for outdoor sports such as rock-climbing and rafting. ◯

7 Give three problems caused by tourism in the French Alps.

a) ....

b) ....

c) ....

8 Fill in the missing words to complete the following sentences.

a) Glaciers in the Alps have retreated by .... in the last 150 years. Snowfall is

becoming increasingly unreliable, so resorts are using .... snow, which is

having an impact on local .... supplies.

b) A 'glacial outburst' could happen when a .... on top of a glacier breaks
through the ice and flows down a valley.

Glacial Landforms

Corries

1 Which of the following describes a corrie? Tick the correct option.

A A steep-sided ridge at the top of mountain. ◯

B A long thin lake at the bottom of a valley. ◯

C A deep circular hollow at the top of a mountain. ◯

D A sharply pointed mountain summit. ◯

2 The diagram shows a cross-section of a corrie. Match statements **A, B, C, D, E, F** and **G** with the labels **1–7** on the diagram. Enter the appropriate number in the boxes provided.

A Glacier ◯   **B** Lip ◯

C Abrasion ◯   **D** Steep back wall ◯

E Crevasses ◯   **F** Plucking ◯

G Freeze-thaw weathering ◯

Arêtes and Pyramidal Peaks

3 Choose the correct words from the options given to complete the following sentences. (You may use a word more than once.)

corries	arêtes	ridges	summit

a) Arêtes are narrow, steep-sided _____ between two _____ .

The sides and back walls of the _____ are eroded until they meet.

b) A pyramidal peak is a sharp, pointed mountain _____ . Three or more

_____ are enlarged by weathering and erosion until they meet.

◯

Glacial Landforms

Glacial Troughs

4 Circle the correct options in the following sentences.

a) A glacial trough is a wide **V-shaped / U-shaped** valley.

b) As a glacier moves down a valley it widens and deepens it by **erosion / deposition**.

c) The ridges of land that the valley used to meander around are cut off to form **interlocking / truncated** spurs.

d) Smaller glaciers that join a larger glacier are known as **tributary / watershed** glaciers. When they melt they leave behind **hanging / suspended** valleys.

e) As the glacier melts, water can be trapped behind glacial till, creating a **tarn / ribbon lake.**

Moraines

5 Which of the following statements about moraines are true? Tick the correct options.

A Moraine is rock that has been transported and deposited by a glacier.

B Another name for moraine is alluvium.

C Moraine can be as fine as clay or as large as boulders.

D Lateral moraine is found along the sides of valleys.

E Ground moraine is moraine that has been pulverised into a powder by a glacier.

F Medial moraine is found in the middle of a glacier.

G Terminal moraine is found at the end of a glacier and marks the furthest point that the glacier reached.

Drumlins

6 Look at the diagram of a swarm of drumlins. Add an arrow to the diagram to indicate the direction that the glacier was moving when the drumlins were formed.

7 Explain the formation of a drumlin.

....

....

Coastal Processes

Waves

1 Fill in the missing words to complete the following sentences.

The movement of a wave up a beach is known as _____. The movement of a wave

down a beach is known as _____.

2 Circle the correct options in the following sentences.

a) **Destructive / Constructive** waves remove beach material.

b) **Destructive / Constructive** waves are long relative to their height.

c) Constructive waves have a **stronger / weaker** swash compared to backwash.

Erosion

3 The table contains four types of coastal erosion. Match descriptions **A, B, C** and **D** with the terms **1–4** in the table. Enter the appropriate number in the boxes provided.

1	Hydraulic power
2	Abrasion
3	Attrition
4	Solution

A   Wearing away of cliffs by material carried by the sea. ◯

B   Rocks containing calcium carbonate are dissolved by acidic sea water. ◯

C   The force of waves crashing into rocks. ◯

D   Rocks carried by the sea are eroded as they rub against each other. ◯

Transportation

4 Choose the correct words from the options given to complete the following sentences.

| Traction | Saltation | Suspension | Solution |

a) _____ is beach material being bounced along by the action of waves.

b) _____ is beach material being rolled along by the action of waves.

c) _____ is where rock is dissolved and then transported in the sea water. ◻

d) _____ is beach material being carried in the sea water.

Coastal Processes

Deposition

Revision Guide Reference: Page 40–41

5 Give three sources of beach material.

a) ....

b) ....

c) ....

6 On a beach, where would you find the largest material?

....

Mass Movement and Weathering

7 Which of the following statements about mass movement and weathering are true? Tick the correct options.

A Mass movement is the movement of rock and soil down a slope. ⬭

B Mass movement is often triggered by dry spells of weather. ⬭

C Sliding is where blocks of material move rapidly down a slope. ⬭

D Slumping is where blocks of material slip along a curved plane. ⬭

E A 'toe' is the exposed surface of rock left behind after slumping. ⬭

Coastal Flooding – Case Study

8 a) Give two reasons why global warming causes sea levels to rise.

i) ....

ii) ....

b) Describe the effects of sea level rise on each of the following for the nation of Kiribati in the Pacific Ocean:

i) Homes: ....

ii) Water supply: ....

iii) Exports: ....

Coastal Landforms

Headlands and Bays

1 Draw lines between the boxes to match the types of rock to the features they form.

Harder rock		Forms bays
Softer rock		Forms headlands

2 a) Give an example of a type of hard rock. ....

b) Give an example of a type of soft rock. ....

Cliffs and Wave-cut Platforms

3 Circle the correct options in the following sentences.

a) A wave-cut platform is a flat area of rock found at the **face / base** of a sea cliff.

b) **Hydraulic power / Attrition** undercuts cliffs to form a wave-cut notch.

c) Over time the cliff collapses and retreats, forming a **raised beach / wave-cut platform**.

Caves, Arches, Stacks and Stumps

4 The table contains six features of an eroded headland. Match descriptions **A, B, C, D, E** and **F** with the terms **1–6** in the table. Enter the appropriate number in the boxes provided.

1	Fault
2	Crack
3	Cave
4	Arch
5	Stack
6	Stump

A The remains of a stack that has been destroyed by erosion.

B Where a fault has started to be eroded by wave action.

C An opening in a headland where a crack has been eroded.

D A natural weakness in a rock.

E Where erosion has broken completely through a headland.

F A pillar of rock that has become separated from a headland.

Beaches, Spits and Bars

5 Which of the following statements about beaches are true? Tick the correct options.

A Beaches are formed by deposition. ⬚

B Beaches are made from sand and pebbles. ⬚

C Beach material is deposited by destructive waves. ⬚

D Longshore drift is a process that moves beach material along a coast. ⬚

E The smaller the beach material, the steeper the gradient of the beach. ⬚

6 Choose the correct words from the options given to complete the following sentences.

deposited   secondary   beach   transported   prevailing   salt

a) A spit is a curved .... that extends into the sea wherever the coast changes

direction. Beach material is .... by longshore drift. The direction of longshore

drift is determined by the .... wind.

b) The hooked end of a spit is caused whenever a .... wind blows.

Silt is often .... in the sheltered area behind a spit to form a

.... marsh.

7 Which of the following describes a bar? Tick the correct option.

A A ridge of material that stretches out to sea from a beach. ⬚

B A ridge of material that connects an island to the mainland. ⬚

C A ridge of material that runs along the length of a beach. ⬚

D A ridge of material deposited parallel to a coastline. ⬚

8 What is a barrier beach?

....

....

Coastal Management

Coastal Management Strategies

1 Circle the correct options in the following sentences.

   a) The use of structures or machinery to control natural processes is known as **hard / soft** engineering.

   b) Working with the environment to modify or prepare for natural processes is known as **hard / soft** engineering.

2 The table contains six coastal management strategies. Match descriptions **A, B, C, D, E** and **F** with the terms **1–6** in the table. Enter the appropriate number in the boxes provided.

1	Sea walls
2	Groynes
3	Rock armour
4	Beach nourishment
5	Dune regeneration
6	Managed retreat

   A   Boulders of resistant rock placed at the base of cliffs.

   B   Allowing the sea to erode and flood some locations, forming natural coastal defences.

   C   Wooden barriers built across a beach to trap material carried by longshore drift.

   D   Replacing eroded beach material by dredging or importing sand.

   E   Designed to reflect wave energy and prevent flooding.

   F   Fencing off areas of a beach to allow vegetation to grow, which will stabilise the sand.

Coastal Management – Case Study

3 Which of the following statements about the Holderness coast are true? Tick the correct options.

   A   The Holderness coast is on the west coast of the UK.

   B   The cliffs are made from glacial till known as boulder clay.

   C   The cliffs are resistant to coastal erosion.

   D   The cliffs are retreating at a rate of 10 metres a year in some locations.

4 Describe a positive effect and a negative effect of the coastal management strategy at Mappleton.

   a) Positive effect: _____

   b) Negative effect: _____

Coastal Environments – Case Study

5 Choose the correct words from the options given to complete the following sentences.

| vegetation | marram | couch | wind | ecosystem |

a) Studland Beach is a beach and sand dune _____ on the Isle of Purbeck in Dorset.

b) Sand dunes are formed where _____ traps sand that has been transported by

_____ from a beach.

c) Vegetation has adapted to the coastal environment by being tolerant to salt, such as

_____ grass, or having long roots to reach fresh water, such as

_____ grass.

6 What is the benefit of rabbits living in sand dune ecosystems like the one at Studland Beach?

_____

7 Which of the following organisations is responsible for managing Studland Beach? Tick the correct option.

A English Heritage ◯

B The Environment Agency ◯

C The National Trust ◯

D Council for the Protection of Rural England ◯

E DEFRA ◯

8 Write down three ways in which the impact of tourism at Studland Beach has been managed.

a) _____

b) _____

c) _____

Map Skills

To answer the questions on these two pages, refer to the Ordnance Survey map extract on page 88.

Direction and Scale

1 Circle the correct options in the following sentences.

a) Ducklington is **north / south** of Witney.

b) Minster Lovell is **east / west** of Witney.

c) Brize Norton is **south-east / south-west** of Witney.

2 Fill in the missing words to complete the following sentences.

a) On a 1:50 000 map, 2cm on the map is equal to _____ in reality.

b) On a 1:25 000 map, 2cm on the map is equal to _____ in reality.

c) On any Ordnance Survey map, the distance from one side of a grid square to the other side is

always _____.

d) On any Ordnance Survey map, the area covered by one grid square is always _____

Grid References

3 Choose the correct words from the options given to complete the following sentences.

        industrial estate   **golf course**   **sewage works**

Land use at the fringe of Witney includes the _____ _____

at 3408, the _____ _____ at 3209 and the

_____ _____ at 3310.

4 The table below contains four places that might be visited by tourists in and around Witney. Match the six-figure grid references **A, B, C** and **D** with the places **1–4** in the table. Enter the appropriate number in the boxes provided.

1	Museum
2	School
3	Hotel
4	Historic remains

A 354093 ◯   **B** 326114 ◯   **C** 353085 ◯   **D** 355100 ◯

Relief

5 Which of the following statements about how relief is shown on a map are true? Tick the correct options.

    A Contour lines are thin brown lines that show the height of the land in metres above sea level. ◯

    B The closer together the contour lines are, the flatter the slope is. ◯

    C A spot height is marked as a dot with the height in metres next to it. ◯

    D A triangulation point is shown as a red triangle with the height in metres next to it. ◯

    E The interval between contour lines is usually 10 metres. ◯

6 a) What is the height of the triangulation point at 316154 in Leafield? ....

    b) What is the height of the spot height at 325065 on the A4095 at Lew? ....

    c) Why was Brize Norton Airfield built in grid squares 2905 and 2906?

    ....

    d) Describe the shape of the land that the River Windrush flows through from grid square 2911 to 3311.

    ....

    ....

Measuring Distance

7 a) How far is it from the PH in Minster Lovell at 315111 along the B4477 due south to the junction with the A40 at 310095? Tick the correct option.

    A 1km ◯   **B** 1.5km ◯

    C 2km ◯   **D** 3km ◯

    b) How far is it from the public telephone at Asthall Leigh at 308124 along the Roman Road in a north-easterly direction to the B4022 at 344145? Tick the correct option.

    A 1km ◯   **B** 2km ◯

    C 4km ◯   **D** 8km ◯

    c) What is the approximate area of woodland covered by Hens Grove and Stockley Copse (Grid Square 2913)? Tick the correct option.

    A $1m^2$ ◯   **B** $10m^2$ ◯

    C $100m^2$ ◯   **D** $1km^2$ ◯

Population Change

Global Population Growth

1 Fill in the missing numbers to complete the following sentences.

Global population reached _____ billion people in 1800. The population in 2009 is

estimated to be _____ billion people. Global population is expected to continue to

grow until it reaches a peak of _____ billion people in 2300.

2 Circle the correct options in the following sentences.

a) The rapid growth in population is known as the population **detonation / explosion**.

b) The increase in population can be explained by improvements in **healthcare / contraception**.

c) 95% of the growth is occurring in **richer / poorer** countries.

3 **a)** What is meant by birth rate?

_____

b) What is meant by death rate?

_____

c) What is the difference between the birth rate and the death rate known as?

_____

The Demographic Transition Model

4 Which of the following describes the demographic transition model? Tick the correct option.

A It shows how changes in birth and death rates affect population growth. ◯

B It shows the structure of a population in terms of age, sex and life expectancy. ◯

C It shows the changing population density in the world. ◯

D It shows how many people are moving from one country to another at any one time. ◯

The Demographic Transition Model (cont.)

5 The table contains the five stages of the demographic transition model. Match descriptions **A, B, C, D** and **E** with the terms **1–5** in the table. Enter the appropriate number in the boxes provided.

1	High Stationary
2	Early Expanding
3	Late Expanding
4	Low Stationary
5	Declining

A High birth rates, falling death rates, population grows rapidly. ◯

B Low birth rates, low death rates, population is stable. ◯

C Falling birth rates, low death rates, population declines slowly. ◯

D Falling birth rates, low death rates, population growth slows. ◯

E High birth rates, high death rates, population grows slowly. ◯

6 a) Give two reasons why death rates fall as a country becomes richer.

i) ....

ii) ....

b) Give two reasons why birth rates fall as a country becomes richer.

i) ....

ii) ....

Population Structures

7 Which of the following statements about population structures are true? Tick the correct options.

A Population pyramids tell you the make-up of a population in terms of age, sex and life expectancy. ◯

B Population is divided into 5-year age groups. ◯

C Females are shown on the left and males on the right. ◯

D India has a high birth rate and a high death rate. ◯

E Japan has a rising birth rate. ◯

Population Strategies

Impacts of Population Change

1 Fill in the missing words to complete the following sentences.

    a) As population grows there will be increasing demand for resources such as

    _____ and _____. It's possible that demand will exceed

    _____.

    b) Increasing demands on services will put pressure on _____ and

    _____; additional _____ will need to be built to accommodate

    everyone. The environment will become threatened by increasing _____ at all
    scales.

Population Policies

2 Give three reasons why people in some countries have large families.

    a) _____

    b) _____

    c) _____

Case Study: China

3 What was the population of China in 2008? Tick the correct option.

    A 1 billion ◯   **B** 1.1 billion ◯

    C 1.3 billion ◯   **D** 1.5 billion ◯

4 Which of the following statements about China's population policy are true? Tick the correct options.

    A A one-child policy was introduced in 1979. ◯

    B People need permission from the government to marry or have children. ◯

    C Free healthcare and education is available to families with only one child. ◯

    D China has strict anti-abortion laws. ◯

    E If families have more than one child they have to give it up for adoption. ◯

    F In rural areas families may be allowed a second child if their first is a girl. ◯

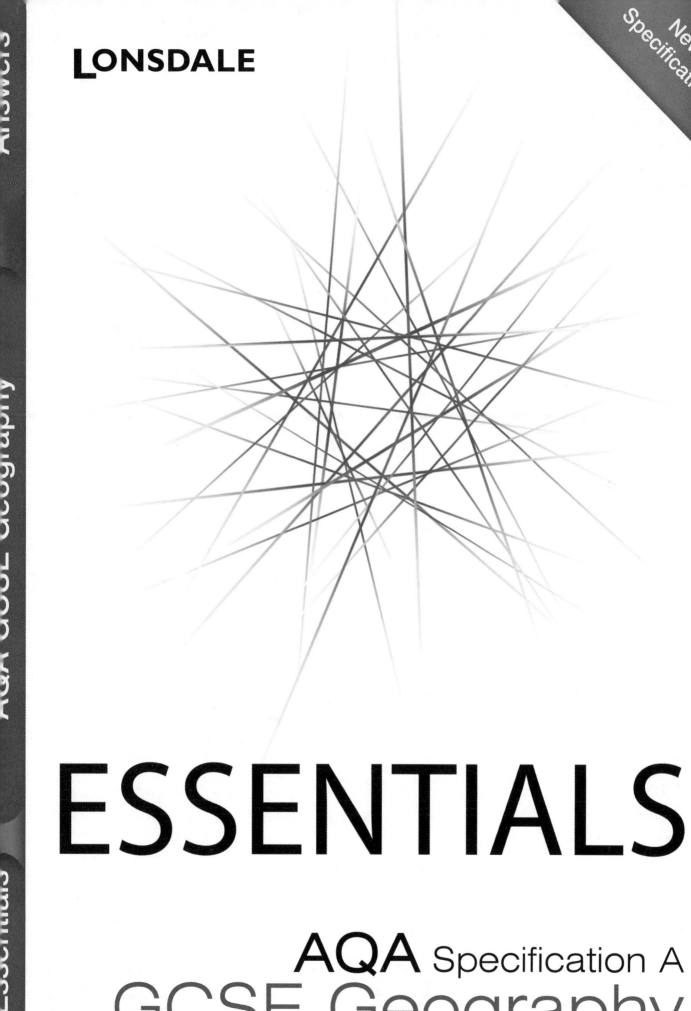

LONSDALE

ESSENTIALS

AQA Specification A
GCSE Geography
Workbook Answers

Geography Essentials Workbook Answers

Page 4

1. **a)** 7
   b) basalt
   c) 25–90km
   d) denser

2. A2; B1; C4; D3

3. destructive; anticlines; synclines

4. A large downfold in the Earth's crust that fills with sediment and may eventually form a new range of fold mountains.

Page 5

5. B and C

6. apart; shield; mid-ocean ridge

7. The line between two plates on a conservative plate margin.

8. The plates lock together, pressure builds up and major earthquakes are caused when the fault line breaks.

Page 6

1. An especially hot area in the mantle.

2. A and D

3. Mauna Loa

4. destructive; acid; steep; Mt St Helens (Washington, USA)

5. B

6. **a)** 74 000 years ago.
   b) Toba / Indonesia
   c) Ash would kill people and would trigger global cooling.

Page 7

7. A4; B2; C3; D1

8. destructive; ash; pyroclastic flows; composite

9. **a)–b) Any two from:** Pyroclastic flow burned buildings and trees; Ash buried more than two thirds of the island; 60% of housing was destroyed.

Page 8

1. A2; B4; C1; D3

2. **a)** seismograph
   b) Mercalli

3. Muzaffarabad

4. 7.6; 73 000; 3 million

5. A and B

Page 9

6. D

7. **a)** 1600km
   b) 12 metres
   c) 500mph
   d) 30 metres

8. 1.7 million; £7.5 billion; warning system

9. **a)–c) In any order:** Create hazard maps; Make emergency plans and prepare supplies; Earthquake drills.

Page 10

1. A4; B2; C1; D3

2. molten rock

3. **a)** intrusive
   b) dark
   c) coarse

4. A3; B1; C4; D2

5. Metamorphic rocks are formed from sedimentary and igneous rocks that have been put under immense heat and pressure.

Page 11

6. Marble – Chalk and limestone
   Slate – Clay and mudstone
   Quartzite – Sandstone

7. The roots of trees and plants are able to break apart rocks that they grow through. The roots also release humic acid, which dissolves some rocks.

8. B

9. **a)** cracks
   b) expands
   c) melts

10. During the day the sun heats rocks and they expand. At night the rocks cool down and contract. Over time, repeated heating and cooling causes the outer layers of the rocks to break off. This is known as exfoliation.

Page 12

1. D

2. A, C, D and F

3. **a) Any suitable answer, for example:** Grazing; China clay (paint / pottery / paper); Tourism.
   b) Any suitable answer, for example: Infertile land; Marshes.

4. A4; B3; C2; D1

Page 13

5. A2; B1; C5; D3; E4

6. **a)–b) Any suitable answers, for example:** Landscape ruined; Dust and noise.

7. **a)** jobs; materials
   b) lake; nature reserve

Page 14

1. A

2. maritime

3. **a)** July
   b) west

4. A4; B1; C5; D2; E3

Page 15

5. **a)** pressure
   b) front
   c) warm; cold
   d) precipitation

6. **a)** Cold, clouds are high up, no rain.
   b) Warm, no clouds, no rain.
   c) Cold, cloudy, rain.

7. A, B and D

Page 16

1. A2; B1; C3

2. **a)** 1000–1400
   b) 500 years
   c) 1°C

3. C

4. Methane; Sunspots

Page 17

5. **a)** 145 million
   b) i)–ii) **In any order:** 4; 5
   c) Malaria
   d) 250 million
   e) 30%

6. **a)** Kyoto Protocol
   b) carbon credits
   c) alternative / renewable
   d) congestion charging
   e) conserving energy

Page 18

1. A, B, E and F

2. A3; B2; C5; D4; E1

3. A4; B1; C3; D2

4. They responded too slowly.

Page 19

5. **a)** 95%
   b) 133 000
   c) 1 million
   d) $10 billion

6. The country's military rulers initially refused to allow foreign aid into the country. Burma is a poor country with limited resources available to respond to natural disasters.

7. D

8. **a)** 55 000
   b) 300 000
   c) $2 billion

Page 20

1. A community of trees, plants, animals and insects living together in a particular environment.

2. A2; B4; C3; D1

3. **a)** chain
   b) photosynthesis
   c) producers
   d) consumers

4. **a)** species
   b) agriculture

Page 21

5. C

6. **a)** mild; warm
   b) 1000mm
   c) brown soil

7. **a)** i)–iii) **Any three suitable answers, for example:** Oak; Ash; Chestnut; Beech; Elm
   b) i)–ii) **Any two suitable answers, for example:** Hazel; Holly; Hawthorn
   c) **Any suitable answer, for example:** Grasses; Bracken; Bluebells

8. A

Page 22

1. Equatorial areas including Brazil, West Africa and South East Asia.

2. C

3. **a)** buttress
   b) waxy
   c) shallow

4. **Any suitable answer, for example:** Farming; Logging; Road building; Mineral extraction; Population pressure.

5. A4; B1; C3; D2

Page 23

6. A, C and E

7. **a)** Tropics of Cancer and Capricorn.
   b) high
   c) xerophytic
   d) succulents

8. **Any suitable answer, for example:** To visit the casinos; For entertainment; For the sunny weather.

9. C

Page 24

1. cubic metres per second; increases

2. A5; B1; C4; D6; E3; F2

3. m/sec; increases; friction

4. **a)** abrasion
   b) action
   c) corrosion

Page 25

5. A4; B3; C1; D2

6. A and D

7. B

8. Upper – Vertical erosion, steep valley sides. Bedload: boulders, stones.
   Middle – Vertical and lateral erosion, gently sloping valley sides. Bedload: pebbles, stones.
   Lower – Lateral erosion, wide valley floor. Bedload: sand, silt, clay.

Page 26

1. A3; B5; C6; D2; E4; F1

2. A1; B5; C3; D4; E7; F6; G2

3. **a)** On the outside of a meander.
   b) On the inside of a meander.

Page 27

4. breaks; deposition; horseshoe

5. **Any suitable answer, for example:**
   False River, Louisiana.

6. C

7. deposited; largest; flooding

8. **Any suitable answer, for example:**
   Grant River, Wisconsin, USA.

Page 28

1. B, C, E and F

2. **a)** 2004; 77; 2
   b) 300
   c) 58; 100

3. **a)** 2000
   b) Four weeks.
   c) Tropical cyclone.
   d) 400
   e) 1 million

Page 29

4. hard; increase; soft; zoning

5. **a)–c) In any order:** Reservoirs; Rivers; Aquifers

6. B and D

7. **a)** Oxfordshire; four
   b) London; farmland; leaky pipes

Page 30

1. A period of long-term cooling of the Earth's atmosphere.

2. **a)** 2 million
   b) inter-glacials
   c) 10 000
   d) 30%
   e) high

3. A4; B3; C2; D5; E1

4. Ablation – The decrease in glacial ice from melting and evaporation.
   Glacial budget – The balance between accumulation and ablation.
   Accumulation – The increase in glacial ice from snow and precipitation.

Page 31

5. A4; B2; C3; D1

6. A, C and D

7. **a)–c) Any three from:** Pressure on the fragile environment; Deforestation increases the risk of avalanches; Footpath erosion; Seasonal unemployment.

8. **a)** 30%; artificial; water
   b) lake

Page 32

1. C

2. A6; B4; C5; D1; E3; F7; G2

3. **a)** ridges; corries; corries
   b) summit; corries

Page 33

4. **a)** U-shaped
   b) erosion
   c) truncated
   d) tributary; hanging
   e) ribbon lake

5. A, C, D, F and G

6. The arrow should be pointing to the right.

7. Till is deposited under a glacier. The glacier shapes the till as it moves. The sharp end of a drumlin points in the direction that the glacier was moving in.

Page 34

1. swash; backwash

2. **a)** Destructive
   b) Constructive
   c) stronger

3. A2; B4; C1; D3

4. **a)** Saltation
   b) Traction
   c) Solution
   d) Suspension

Page 35

5. **a)–c) In any order:** Eroded cliffs; Sediment banks; River bedload.

6. At the back of the beach.

7. A, C and D

8. **a) i)–ii) In any order:** As the sea warms up it expands (thermal expansion); As glaciers and ice sheets melt sea levels will rise.
   b) i) Homes: People have to rebuild their homes further from the sea.
   ii) Water supply: Fresh water is being contaminated by salt water.
   iii) Exports: Coconut trees are being killed by salt water.

Page 36

1. Harder rock – Forms headlands
   Softer rock – Forms bays

2. **a) Any suitable answer, for example:** Limestone.
   b) Any suitable answer, for example: Clay.

3. **a)** base
   b) hydraulic power
   c) wave-cut platform

4. A6; B2; C3; D1; E4; F5

Page 37

5. A, B and D

6. **a)** beach; transported; prevailing
   b) secondary; deposited; salt

7. D

8. A large bar formed when sea levels rose at the end of the last ice age.

Page 38

1. **a)** hard
   b) soft

2. A3; B6; C2; D4; E1; F5

3. B and D

4. **a)** Erosion has been prevented, protecting homes and roads.
   b) Erosion to the South has increased, so farmland and houses have been lost.

Page 39

5. **a)** ecosystem
   b) vegetation; wind
   c) couch; marram

6. Their droppings add organic material.

7. C

8. **a)–c) In any order:** Charging for car parking; Litter bins on the beach; Maintaining footpaths.

Page 40

1. **a)** South
   b) West
   c) South West

2. **a)** 1km
   b) 0.5km
   c) 1km
   d) 1km^2

3. sewage works; golf course; industrial estate

4. A2; B4; C3; D1

Page 41

5. A, C and E

6. **a)** 195m
   b) 79m
   c) Low lying flat land, 70m–80m contour lines.
   d) A river valley with a floodplain, 100m at the top of the valley sloping to 80m at the floodplain.

7. **a)** B
   b) C
   c) D

Page 42

1. 1; 6.9; 9

2. **a)** explosion
   b) healthcare
   c) poorer

3. **a)** The number of babies born per 1000 people.
   b) The number of deaths per 1000 people.
   c) Natural increase / decrease.

4. A

Page 43

5. A2; B4; C5; D3; E1

6. **a) i)–ii) Any suitable answers, for example:** Improvements in healthcare; Improvements in diet.
   b) i)–ii) Any suitable answers, for example: Fewer children are needed for labour; Women choose education and careers instead of having children.

7. A, B and D

Page 44

1. **a) The first two spaces can be filled with any two of the following in any order:** water / energy / land / food. **The second space must be filled with:** supply
   b) healthcare; education; housing; pollution **OR** education; healthcare; housing; pollution

2. **a)–c) Any three from:** Children are needed to work on a farm; Children are needed to contribute to the family income; People rely on children to look after them when they get older; High rates of infant mortality mean that families have several children in case some die young.

3. C

4. A, B, C and F

Page 45

5. **a)** 98; contraception; abortion; 21
   b) 30; youthful

6. C

7. 15; 3

8. **a)–c) In any order:** Encourage immigration of skilled workers; Raise the retirement age; Build more sheltered housing / hospitals, and fewer schools.

Page 46

1. A, B and D

2. **a) i)–iii) Any three from:** Lack of jobs; Poor housing; Lack of healthcare; War; Natural disasters.
   b) C

3. By encouraging some migrants and discouraging others, usually through strict border controls and immigration policies.

4. jobs; wages; services; jobs; wages; services

5. D

6. push; pull

Page 47

7. **a) i)–iv) Any four from:** Poland; Lithuania; Latvia; Estonia; Hungary; Slovenia; Slovakia; Czech Republic.
   b) i) Any suitable answer, for example: Skilled workers are brought into the UK, who pay UK taxes.
   ii) Any suitable answer, for example: Resentment from UK workers; Lack of skilled workers in countries of origin.
   c) 1 million

8. A, C, D and E

Page 48

1. B

2. 80; 300; poorer

3. A and B

4. A4; B3; C1; D2

Page 49

5. **a)** cities
   b) poor
   c) employment
   d) higher
   e) 60%

6. C

7. **a)** shops; offices; centre
   b) squatter settlements; outskirts
   c) self-built; permanent

Page 50

1. Where people can live and work in such a way that doesn't prevent people in the future from living and working.

2. shortage; high; average

3. A, C and D

4. D

5. **a)–b) Any suitable answers, for example:** Competition from out-of-town shopping centres; Old and rundown buildings.

Page 51

6. **a)** Through redevelopment, knocking down old shopping centres and building new ones.
   b) Pedestrianisation.

7. cities; cluster together

8. A and B

9. Britain's greenest city.

10. **a)** 20%
    b) plant trees
    c) free
    d) discounted
    e) expanded

Page 52

1. A, B, C and D

2. **a)** toxic; carbon monoxide; sulphur dioxide
   b) banned; fined

3. **a)–b) In any order:** Domestic waste; Industrial waste.

4. **a)–b) Any two from:** Poor water supply; No sewage system; No waste collection; Risk of disease; No healthcare; Unemployment; Violent crime.

5. Barrio – Latin America
   Favela – Brazil
   Bustee – India
   Kampong – South East Asia

Page 53

6. A3; B4; C1; D2

7. D

8. **a)** Bairio
   b) widened; bricks; water
   c) adults

1. A

2. a)–b) **Any two from:** Modern housing estates; Industrial estates; Business parks; Out-of-town shopping centres; Farms and woodland.

3. developers; pleasant; space; cheap; roads; commuters

4. B and C

5. It has increased property prices and led to friction with locals.

6. a)–b) **Any two from:** 1 post office; 1 bakery; 2 banks; 2 greengrocers; 2 butchers.

7. a) A decline in population in rural areas.
   b) i)–ii) **In any order:** (Fewer young people mean) fewer babies are being born in rural areas; Rural to urban migration (for work and leisure).

8. B

9. A, B and D

1. D

2. a) flat
   b) loam
   c) low
   d) ripen crops

3. companies; 200; chemicals; agribusiness

4. A, C and E

5. a) Farming with no artificial fertilizers, herbicides or pesticides, and no intensive animal rearing.
   b) 4%

6. A5; B4; C3; D1; E2

7. hedgerows; machinery; habitats; decrease

8. A1; B3; C2

1. a) subsistence
   b) cultivation; 4–5
   c) ash
   d) fertility
   e) 50

2. A3; B1; C4; D2

3. A, B and D

4. a)–b) **Any two from:** Fertile land is taken up, which could be used by local farmers; Soil erosion is caused by machinery; A single crop is vulnerable to pests and diseases.

5. C

6. high; fertilizers; machinery; 200%; two; afford; pollution

7. a) The artificial application of water to assist in the growing of crops.
   b) Because rainfall is seasonal.

8. Intercropping – Growing a variety of plants to maximise productivity
   Food storage – Prevention of crop loss by rodents, insects or disease
   Natural fertilizer – Leguminous crops that add nitrogen to the soil

1. A7; B3; C1; D8; E4; F5; G6; H2

2. A, B and D

3. A8; B7; C4; D3; E5; F2; G1; H6

4. A country that's becoming richer, that has experienced recent rapid economic growth.

5. a) well-being; health; happiness **OR** well-being; happiness; health
   b) goods; services **OR** services; goods
   c) perception

1. infrastructure; resources; rebuilding; education

2. a) lower
   b) higher

3. 1.2; death; healthcare

4. A3; B1; C2

5. a) A social, political and economic union of 27 countries in Europe.
   b) i)–iii) **Any three from:** UK; France; Ireland; Spain; Denmark; Netherlands; Austria
   c) i)–iii) **Any three from:** Estonia; Latvia; Lithuania; Poland; Hungary; Bulgaria; Romania

6. B, C, E and G

7. B

8. a) Common Agricultural
   b) Regional Aid

1. A, B and C

2. tariffs; quotas; import

3. A2; B5; C4; D3; E1

4. Trade that makes sure that the people who provide the raw materials receive a fair price for their produce and work without exploitation.

5. a)–b) **Any two from:** Minimum wages; Safe working conditions; Restrictions on child labour; Environmental protection; Improved healthcare and education.

6. a) richer; poorer
   b) bilateral
   c) multi-lateral
   d) Short-term
   e) long-term
   f) Conservation swap

7. B, C and F

1. economic; 1950s; interdependent

2. A, C and D

3. a) A company that operates in more than one country.
   b) In rich countries like the UK, USA and Japan.
   c) i)–iii) **In any order:** Jobs are provided; Taxes are paid, which can be re-invested; Technology and skills are passed on.
   d) i)–iii) **In any order:** Profits go to the home nation of the transnational corporation; Wages can be very low; Farming and manufacturing can cause environmental damage.

4. A5; B1; C4; D3; E2

5. A decline in traditional manufacturing industry involving the loss of jobs and ultimately the loss of output.

6. a) lower; cheaper
   b) cheaper
   c) cheapest
   d) long hours
   e) illegal
   f) transnational corporations

1. A

2. a) 6.8 billion
   b) 9 billion

3. B, C and E

4. a) pollution; fossil; nuclear; conflict
   b) rise
   c) low; prices / bills

5. a) present; future
   b) Japan; 1997

6. A4; B2; C5; D1; E3

7. Local – Putting waste in a recycling bin is an effective way for everyone to save energy.
   National – Governments can introduce policies to encourage people to save energy.
   Global – Governments can sell their quota of greenhouse gas emissions to other countries.

1. a) 2050
   b) 75
   c) processed
   d) industrialised

2. B, C, D, E, G and H

3. a) The distance food has travelled from where it was produced to where it's consumed.
   b) The amount of CO_2 emitted in food production, processing, storage and transportation.

Page 71

4. a) Farming to feed your family.
 b) They believe that they'll make more money growing cash crops.
 c) Fair trade guarantees a fair price and working conditions for families, as well as helping their families with housing, healthcare and education.
 d) 3 million

5. a) Irrigation
 b) conflict; groundwater
 c) 2025

6. 200; taxes; costs; supermarkets

Page 72

1. pleasure; night; months

2. A, B, D, F and G

3. A2; B1; C3; D1; E2; F3

Page 73

4. a) million
 b) richer
 c) the USA

5. A1; B4; C5; D3; E6; F2

6. Discovery – Small numbers of adventurous visitors; few facilities.
 Development – Increasing visitor numbers; facilities being built.
 Consolidation – Large numbers of visitors; facilities owned by large companies.
 Stagnation – Visitor numbers peak; the resort is no longer fashionable.
 Decline – Numbers decline as tourists seek new destinations.
 Rejuvenation – Investment and advertising attracts new visitors.

Page 74

1. a) 5
 b) 2.1
 c) 30; 15

2. a)–c) In any order: Terrorist attacks in New York; Invasion of Iraq; High value of the pound against the dollar.

3. a) National Park; 14
 b) lakes / mountains; mountains / lakes; hill-walking / mountain-biking; mountain-biking / hill-walking **(or any other suitable answer)**; Ambleside / Keswick **(or any other suitable answer)**; Keswick / Ambleside **(or any other suitable answer)**

4. a)–c) In any order: Encouraging more visitors from ethnic minorities; Persuading people to use alternative forms of transport; Making tourism more sustainable.

Page 75

5. A, C and D

6. a) coastal
 b) sandy beaches
 c) medieval walls
 d) tourists

7. A, B, D, F and G

Page 76

1. a) 1970s
 b) air travel
 c) package holidays

2. poor; East; 1990s; scenery; National Parks

3. A, C and E

4. a) current; future
 b) Lamu; local; palm trees; buildings

Page 77

5. B

6. a) 1958
 b) 30 000
 c) Between April and October.
 d) For the scenery, wildlife, remoteness / extreme environment.
 e) i)–iii) Any suitable answers, for example: Money goes to conservation; Visitors want to protect Antarctica – and spread the word; Historic sites have been protected.
 f) i)–iii) Any suitable answers, for example: Penguin colonies are disturbed; Tourists trample slow-growing plants; Tourist ships are getting larger, and there's a risk of oil spill.

7. a) 100
 b) souvenirs
 c) discharge waste

Answers to Exam-style Questions

Page 78

1 (a) (i) The plates move apart. The magma in the mantle is under pressure and lava wells up in the fault between the plates.

1 (a) (ii) Shield volcanoes are lower, less steep and have gentler eruptions than composite volcanoes. The lava is 'basic', runny and flows long distances compared to composite volcano lava.

1 (b) (i) Example: 2004 Asian Tsunami; Causes: Undersea earthquake of Richter Scale 9.2, destructive plate margin on Indo-Australian / Eurasian plates, 1600km-long fault, sea bed raised by 12 metres. The resulting tsunamis were 500mph and 30m high. Effects: 11 countries were affected including Indonesia, Sri Lanka, India and Thailand, 1.7 million people were made homeless, 230 000 people were confirmed dead or missing.

Page 79

2 (a) (i) The highest rainfall is to the West; the lowest rainfall is to the East. Fort William has the most rainfall; London has the least rainfall.

2 (a) (ii) Keswick has relief rainfall; warm moist air from the Atlantic rises over the Lake District, then cools, condenses and falls as rain. As the now dry air moves east towards Newcastle, it sinks and warms up so rain is unlikely to fall.

2 (b) (i) People are affected directly by hurricane force winds, or indirectly by flooding caused by storm surges, high winds and heavy rain. Homes and property can be damaged or destroyed, crops and livestock lost, industry / employment ruined, and people are injured or killed, or suffer from lack of hygiene / acquire illnesses.

2 (c) (i) Example: Hurricane Katrina, USA, 2005. Immediate responses: 1 million people evacuated, shelter provided in sports stadium, police and military rescued people and restored law and order. Long-term responses: Housing assistance was given to 700 000 people, areas of New Orleans were completely rebuilt. Cyclone Nargis, Burma, 2008. Immediate responses: Foreign aid was refused, emergency services were inadequate, hospitals were overwhelmed. Long-term responses: Eventually food, water and medicine donated by other countries were allowed into Burma. Many communities received little or no aid and had to cope with the disaster on their own.

Page 80

3 (a) (i) On the tropics of Cancer and Capricorn ($23\frac{1}{2}°$ North or South of the Equator), on the West coast of continents or in the middle of continents.

3 (a) (ii) It can survive with very little water (xerophytic), waxy leaves / rolled leaves / deep roots. It can store water (succulents), e.g. saguaro cactus, prickly pear cactus, spinifex grass.

3 (b) (i) Hot weather, sun all year round, no rain, attractive landscape, interesting vegetation / wildlife.

3 (b) (ii) It can get too hot during the day and too cold during the night, and there are water shortages (water needs to be pumped from elsewhere) and dust storms

Page 81

4 (a) (i) A – Confluence; B – Ox-bow lake; C – Meander; D – Delta/Mouth

4 (a) (ii) When a river floods, it is able to transport more sediment. As the river flows over its banks the water slows down, loses energy and deposits the sediment it was transporting. Over time, the sediment deposited on the river banks builds up to form levees.

4 (b) (i) Example: Boscastle, UK, 2004. Physical: Heavy rain (77mm in 2 hours), ground waterlogged from previous rainfall, steep slopes funnelled water into Boscastle, confluence of River Jordan and River Valency. Human: Building on the flood plain, river channel narrowed due to building, bridge wasn't wide enough to let floodwater through.

4 (c) (i) Gives immediate protection, can be used for other purposes (e.g. HEP), enables the river to be used for transport / leisure, allows building on the floodplain.

4 (c) (ii) Expensive, has to be maintained / replaced, works against the environment, can cause flooding to be worse, spoils the landscape.

5 (a) (i) Destructive waves have a relatively weak swash and strong backwash; constructive waves have relatively strong swash and weak backwash. Destructive waves erode the coast; constructive waves deposit beach material and build up the coast.

5 (a) (ii) A wave-cut platform is formed from the erosion of the base of cliffs by hydraulic power, abrasion and solution. Over time a wave-cut notch is formed, which gets bigger until the cliff above is no longer able to support itself, so it collapses, retreats and the process begins again. Over time a rock shelf of non-eroded rock is left, which is the wave-cut platform.

5 (b) (i) Example: Mappleton, East Yorkshire, UK. A rock groyne and rock armour rip rap have been used to protect the village of Mappleton. Benefits: The groyne has trapped sand, which has built up a beach that now protects homes and an important coast road; the rock armour further protects the cliff. Costs: Expense of building the defences, trapped sand no longer moves down the coast, long shore drift is removing sand from cliffs South of Mappleton, which is not being replaced, and erosion has now increased as a result of the defences.

1 (a) (i) Lack of clean water supply, electricity, sewage system, waste disposal. High levels of crime. Disease risk high. Lack of healthcare / employment.

1 (a) (ii) Example: Rio de Janeiro, Brazil – Favela Barrio programme. Widened streets, brick houses built, electricity and water supplies installed, sports facilities built, child care centres and adult education facilities provided.

1 (a) (iii) Example: Manchester, UK. Green badge parking scheme, least polluting vehicles get discounted parking. Recycling, composting, green waste collection. Tree planting required for new developments. Bus and tram system. New developments must get 20% of their energy from renewable sources.

2 (a) (i) Any two reasons, for example: Land is undeveloped (green field site); It's adjacent to the A40 for access for customers and deliveries; It's near an existing commercial development; No risk of flooding as not on the flood plain.

2 (b) (i) Any one group, for example: Council for the Protection of Rural England; District Council; Local people; Shop owners in the town centre.

2 (b) (ii) Loss of green space, development would be an example of urban sprawl, increased traffic would cause pollution and congestion, competition with town centre shops could cause the CBD to decline, wildlife would be disturbed and habitats lost.

2 (c) (i) Any one group, for example: Residents of Cogges / Woodgreen / The East of Witney; Owners of the Museum / Leisure centre.

2 (c) (ii) The land is part of the floodplain of the river Windrush. The land provides a 'green wedge' of countryside that runs right to the centre of Witney.

3 (a) (i) A place where a business makes / receives telephone calls to enable customers to manage accounts and services.

3 (a) (ii) Telecommunications and ICT have enabled call centres to be located anywhere in the world, costs of land / labour are lower than in the UK, employees in countries like India are more qualified and are prepared to work harder than British workers.

3 (b) (i) Advantages: employment, improving standards of living, taxes are paid, technology and skills are transferred between countries. **Disadvantages:** profits made go to richer countries, wages on farms and in factories are low, working conditions can be very poor, environmental damage.

3 (c) (i) Population growth, increasing wealth of countries such as India and China, demand for new technology, such as mobile phones.

4 (a) (i) Attractive landscape, quiet / tranquil location, paths and stepping stones to walk along, people like to walk by rivers.

4 (a) (ii) Overcrowding, dropping litter, noise, footpath erosion, traffic congestion and pollution, second home ownership pushes up house prices, seasonal employment means people are out of work during the winter.

4 (b) (i) Example: Lake District National Park, UK. Development concentrated on specific sites (Lake Windermere / Keswick), footpaths have been reinforced, the number of car parks has been increased, visitors from a greater range of ethnic backgrounds have been encouraged, people persuaded to use transport such as walking, cycling or horse riding.

4 (c) (i) Example: Antarctica. **Advantages:** Tour operators contribute money to conservation schemes, visitors become advocates for the protection of Antarctica, conservation of important historical sites such as Scott's Hut from the 1910 British Antarctic Expedition.

4 (c) (ii) Disadvantages: penguin colonies disturbed, tourists trample slow growing, fragile plants (mosses and lichens), litter, an accident could result in a major oil spill.

ACKNOWLEDGEMENTS

The author and publisher are grateful to the copyright holders for permission to use quoted materials and images.

Every effort has been made to trace copyright holders and obtain their permission for the use of copyright material. The authors and publishers will gladly receive information enabling them to rectify any error or omission in subsequent editions. All facts are correct at time of going to press.

Published by Lonsdale
An imprint of HarperCollinsPublishers
77–85 Fulham Palace Road
London W6 8JB

© Lonsdale

ISBN 978-1-906415-70-9

First published 2009

01/030409

British Library Cataloguing in Publication Data.

A CIP record of this book is available from the British Library.

Book concept and development: Helen Jacobs
Commissioning Editor: Rebecca Skinner
Author: Andrew Browne
Project Editor: Katie Galloway
Cover Design: Angela English
Inside Concept Design: Helen Jacobs and Sarah Duxbury
Text Design and Layout: Dragon Digital
Printed and bound in Italy

Mixed Sources
Product group from well-managed forests and other controlled sources
www.fsc.org Cert no. SW-COC-001806
© 1996 Forest Stewardship Council
FSC

FSC is a non-profit international organisation established to promote the responsible management of the world's forests. Products carrying the FSC label are independently certified to assure consumers that they come from forests that are managed to meet the social, economic and ecological needs of present and future generations.

Find out more about HarperCollins and the environment at
www.harpercollins.co.uk/green

Population Strategies

Case Study: Philippines

5 Choose the correct words / numbers from the options given to complete the following sentences.

| youthful | 30 | contraception | 98 | abortion | 21 |

a) The population of the Philippines is _____ million people. Access to

_____ is restricted and _____ is illegal. Half the population is

below _____ years old.

b) The population of the Philippines may double within _____ years. It will become

increasingly difficult for the government to provide food, housing, healthcare and education for this

_____ population.

The Ageing Population of the UK

6 Why does the UK have an ageing population? Tick the correct option.

A The death rate is falling and the birth rate is rising.

B People are having larger families.

C Life expectancy is increasing and people are having fewer children.

D People are choosing to live longer.

7 Fill in the missing words to complete the following sentences.

The number of people over the age of 65 is expected to grow from 9.5 million to

_____ million by 2040. There used to be 22 employed people for every retired

person; by 2024 this will have fallen to fewer than _____ employed people for
every retired person.

8 Suggest three strategies that the UK government may have to adopt in order to cope with the ageing
population.

a) _____

b) _____

c) _____

Population Movement

Push and Pull Factors

1 Which of the following statements about migration are true? Tick the correct options.

A Migration is the movement of people from one place to another. ◯

B Permanent migration is where people decide to stay in the place they have migrated to. ◯

C Internal migration is movement between countries. ◯

D Forced migration is the movement of people against their will. ◯

2 a) Give three examples of push factors.

i) _____ ii) _____ iii) _____

b) Which of the following factors is not a pull factor? Tick the correct option.

A Job opportunities ◯   **B** Free education ◯

C War / persecution ◯   **D** Affordable healthcare ◯

3 How might governments attempt to control international migration?

_____

Rural to Urban Migration

4 Choose the correct words from the options given to complete the following sentences. Use each word twice.

services   jobs   wages

People migrate from the countryside due to few _____, low _____

and limited _____. They are pulled to urban areas in the belief that there will be more

_____, higher _____ and a better quality of _____.

Urban to Rural Migration

5 What is the movement of people from urban areas to the countryside known as? Tick the correct option.

A Urbanisation ◯   **B** Ruralisation ◯

C Commuting ◯   **D** Counter-urbanisation ◯

6 Circle the correct options in the following sentences.

Lack of open space, congestion and crime are examples of urban **push / pull** factors. Attractive countryside and a safer environment are rural **push / pull** factors.

Population Movement

Economic Migration in the EU

Revision Guide Reference: Page 55

7 a) Name four Eastern European countries that joined the EU in 2004.

i) ....

ii) ....

iii) ....

iv) ....

b) Give one advantage and one disadvantage of migrant workers moving from Eastern Europe to the UK.

i) Advantage: ....

....

ii) Disadvantage: ....

....

c) Circle the correct option in the following sentence.

Since 1 May 2004, over **1 million / 2.5 million** people from countries that joined the EU in 2004 have come to the UK to find work.

Refugees

8 Which of the following statements about refugees are true? Tick the correct options.

A Refugees are people who have been forced to leave their homes due to war, persecution or natural disasters.

B There are 3 million refugees in the world.

C Many refugees in Europe have come from Uganda, Vietnam, Afghanistan, Kosovo and Iraq.

D It can be difficult to tell the difference between a genuine refugee and an economic migrant.

E Refugees increase pressure on housing, healthcare and education.

F The world's richest nations take more refugees than the world's poorest nations.

Urbanisation

Urbanisation

1 What does the term 'urbanisation' mean? Tick the correct option.

    A The movement of people from rural areas to urban areas. ◯

    B An increase in the proportion of people living in towns and cities. ◯

    C The movement of people from urban areas to rural areas. ◯

2 Fill in the missing words / numbers to complete the following sentences.

In 1950 there were _____ cities in the world with a population of at least one million.

Today there are over _____ cities with a population of at least one million.

The majority of these cities are found in _____ countries.

Urbanisation in Richer Countries

3 Which of the following statements about urbanisation in the richer world are true? Tick the correct options.

    A Urbanisation in the richer world began in the 19th Century during the Industrial Revolution. ◯

    B People moved from rural areas to urban areas to find employment. ◯

    C Most of the factory workers lived in the suburbs in semi-detached houses. ◯

Urban Land Use in Richer Countries

4 The table contains four terms relating to urban land use. Match descriptions **A, B, C** and **D** with the terms **1–4** in the table. Enter the appropriate number in the boxes provided.

1	Central business district
2	Inner city
3	Suburbs
4	Rural–urban fringe

    A Zone with modern housing estates, business / retail parks and leisure centres. ◯

    B Zone with semi-detached and detached houses, built between the 1930s and now. ◯

    C Zone containing mainly shops, offices, restaurants and entertainment venues. ◯

    D Zone with a mix of 19th-century terraced housing, high-rise council flats and modern apartments. ◯ ▢

Urbanisation in Poorer Countries

5 Circle the correct options in the following sentences.

a) Rural to urban migration has increased the number of people living in **cities / villages**.

b) People often leave the countryside due to **good / poor** harvests.

c) People are attracted to urban areas by **employment / unemployment**.

d) Population growth is often **higher / lower** in poorer countries than in richer countries.

e) It is predicted that **50% / 60%** of the world's population will live in cities by 2025.

Urban Land Use in Poorer Countries

6 Which of the following statements about the quality of housing in cities in poorer countries is true? Tick the correct option.

A   The quality of housing increases from the CBD to the edge. ◯

B   The quality of housing stays the same from the CBD to the edge. ◯

C   The quality of housing decreases from the CBD to the edge. ◯

D   Most good quality housing is found between the CBD and the edge. ◯

7 Choose the correct words from the options given to complete the following sentences.

offices   squatter settlements   permanent   centre

outskirts   self-built   shops

a) The CBD of a city in a poorer country will always contain .... and

.... Large colonial houses and modern apartments will be found close to the

.... , along main roads.

b) Self-built houses made from scrap wood, metal and plastic are known as ....

.... They tend to be found at the .... of cities in poorer

countries.

c) Upgraded squatter settlements refer to .... housing that has become

....

Urban Living in the Richer World

Urban Issues in Richer Countries

1 What is a sustainable environment?

_____

_____

Housing

2 Fill in the missing words to complete the following sentences.

In London there is a _____ of housing, which has led to _____

property prices. People with _____ wages are finding it very hard to buy or rent
somewhere to live.

3 Which of the following statements, describing strategies to solve the housing problem in London, are
true? Tick the correct options.

A New homes are being built on old industrial areas. ◯

B Homes are being built on greenfield sites within London. ◯

C New housing developments have to include low-cost dwellings. ◯

D High-rise council flats have been upgraded to provide homes for teachers and nurses. ◯

E People have to apply for a permit before they're allowed to move to London. ◯

Traffic

4 What is the peak traffic flow during the morning and evening known as? Tick the correct option.

A Commuting ◯   **B** Congestion charge ◯

C Traffic jam ◯   **D** Rush hour ◯

Improving the CBD

5 Give two reasons for the decline of some CBDs.

a) _____

b) _____

Urban Living in the Richer World

Improving the CBD (cont.)

Revision Guide Reference: Page 58–59

6 a) How have cities such as Birmingham and Manchester improved their CBDs?

....

....

b) What is the term for closing a street to traffic in order to make it safer and more pleasant for shoppers?

....

Cultural Mix

7 Circle the correct options in the following sentences.

Migrants are attracted mainly to **cities / villages**. People from the same backgrounds tend to **spread out / cluster together**, which can lead to tension between ethnic groups.

8 Which of the following statements, describing how cities have tried to reduce tension between ethnic groups, are true? Tick the correct options.

A Migrants to the UK have to take a citizenship test. ◯

B Police forces are actively recruiting from ethnic groups that reflect the urban community. ◯

C Housing has been set apart so that people from the same ethnic group can live together. ◯

Sustainable Urban Living – Case Study: Manchester

9 What is the city of Manchester hoping to become?

....

10 Circle the correct options in the following sentences.

a) New building developments have to gain **10% / 20%** of their energy from renewable sources.

b) New building developments are required to **plant trees / use recycled materials**.

c) Buses in the city centre are **subsidised / free**.

d) Drivers of the least polluting cars get **discounted / free** parking.

e) Manchester's tram network is being **dismantled / expanded**.

Urban Living in the Poorer World

Urban Issues in Poorer Countries

© Letts and Lonsdale

1 Which of the following statements about waste in cities in poorer countries are true? Tick the correct options.

A Urban areas produce a lot of waste. ◯

B Many cities in poorer countries don't have a central waste collection service. ◯

C Rotting waste attracts rats, which spread disease. ◯

D Providing wheelie bins and a daily collection is a solution. ◯

E Most waste in poorer cities is buried in landfill sites. ◯

2 Choose the correct words from the options given to complete the following sentences.

banned　　**carbon monoxide**　　**sulphur dioxide**　　**fined**　　**toxic**

a) Air pollution in poorer cities is caused by cars and factories releasing

.... gases. People's health is damaged by .... and acid

rain is caused by ...., which damages buildings.

b) To reduce air pollution, cars without catalytic converters can be .... from cities

and polluting factories can be ....

3 Name two ways in which rivers can be polluted in cities in poorer countries.

a) .... **b)** ....

Squatter Settlements

4 List two problems that people living in shanty towns have to cope with.

a) .... **b)** ....

5 Draw lines between the boxes to match each regional term for a shanty town with the place it is situated.

Barrio	South East Asia
Favela	India
Bustee	Latin America
Kampong	Brazil

 © Letts and Lonsdale

Squatter Settlement Redevelopment

6 The table contains four terms relating to improving squatter settlements. Match descriptions **A, B, C** and **D** with the terms **1–4** in the table. Enter the appropriate number in the boxes provided.

1	Consolidation
2	Self help
3	Site and service
4	Local Authority

A Squatter settlement dwellers are offered plots of land with a water supply, sewerage system and electricity. They can then buy / rent the land and build on it.

B Education, healthcare, waste collection and a police force can be set up to improve the quality of life for people who live in squatter settlements.

C Over several decades, squatter settlements can be transformed into permanent settlements by the residents upgrading their homes.

D The government gives squatters legal ownership of the land and provides cheap building materials to help the people improve their homes.

Case Study: Squatter Settlement Redevelopment

7 What is the population of Rio de Janeiro in Brazil? Tick the correct option.

A 1.5 million

B 5.5 million

C 7.1 million

D 11.7 million

8 Fill in the missing words to complete the following sentences.

a) Favela _____ is a government programme that aims to improve the quality of life in Rio de Janeiro's favelas.

b) Streets have been _____ to improve access to the favelas. Wooden shacks have

been replaced with houses made from _____. Electricity and

_____ supplies have been installed.

c) Education has been provided for children and _____.

Rural UK

Rural–Urban Fringe

1 Which of the following describes the rural–urban fringe? Tick the correct option.

   A   A transition zone between urban areas and the countryside.

   B   A zone of land surrounding an urban area where development is not allowed.

   C   The zone that surrounds the CBD.

   D   An area of countryside protected from development and set aside for people to enjoy.

2 List two land uses that would be found in the rural–urban fringe.

   a) ....   **b)** ....

Urban Sprawl

3 Choose the correct words from the options given to complete the following sentences.

   roads   **space**   **commuters**   **pleasant**   **developers**   **cheap**

   The rural–urban fringe offers several advantages to .... The environment is

   .... , there is .... to build large developments, land tends to

   be .... , and there are good .... making access easy for

   .... from the suburbs.

Commuter Villages

4 Which of the following statements about commuter villages are true? Tick the correct options.

   A   Most people in commuter villages work in agriculture.

   B   Most people in commuter villages work in towns and cities.

   C   New housing developments are common in commuter villages.

   D   Most local shops and pubs have closed down.

Rural Change – Second Homes

5 Why has the growth in ownership of second homes led to problems?

   ....

   ....

Rural Change – Second Homes (cont.)

6 Give two examples of services that Wrotham in Kent has lost in the last 10 years.

a) ....

b) ....

Rural Change – Depopulation and Decline

7 a) What is rural depopulation? ....

b) Give two reasons for rural depopulation.

i) .... **ii)** ....

Sustainable Rural Settlements

8 Which government department is responsible for supporting the rural economy and environment? Tick the correct option.

A Ministry of Defence ⬭

B Department for Environment, Food and Rural Affairs ⬭

C The National Trust ⬭

D The Council for the Protection of Rural England ⬭

E English Heritage ⬭

9 Which of the following statements about how the government is supporting rural areas are true? Tick the correct options.

A The government provides support and advice on issues such as affordable housing and broadband connection. ⬭

B People are being encouraged to buy local produce and use public transport. ⬭

C The government is controlling the number of homes that can be sold as second homes in rural areas. ⬭

D The government has strict planning laws to prevent rural areas from becoming spoilt by development. ⬭

E The government has no plans to allow the building of more houses in rural areas for the next 20 years. ⬭

Farming in the UK

Commercial Farming in East Anglia

1 Which of the following describes commercial farming? Tick the correct option.

    A Farming to feed yourself and your family. ◯

    B Farming that produces a high yield from a small area. ◯

    C Farming that doesn't involve the use of artificial pesticides or herbicides. ◯

    D Farming that produces goods to sell for a profit. ◯

2 Circle the correct options in the following sentences.

    a) East Anglia is good for arable farming because it has **flat / steep** land.

    b) East Anglia has deep, fertile **clay / loam** soil.

    c) Crops grow well in East Anglia due to reliable **high / low** rainfall.

    d) Warm summers in East Anglia help to **ripen crops / break up soil**.

3 Fill in the missing words to complete the following sentences.

Many farms in East Anglia are owned by .... These farms tend to be over

.... hectares in size. The farms use technology and ....

to maximise yields. This type of commercial farming is known as ....

Markets

4 Which of the following statements about supermarkets are true? Tick the correct options.

    A Supermarkets are the main market for agricultural products in the UK. ◯

    B Supermarkets pay farmers the highest prices possible. ◯

    C Supermarkets set farmers strict standards of size, shape and quality of produce. ◯

    D Supermarkets will only buy produce from abroad if they can't source it from farms in the UK first. ◯

    E Supermarkets insist that farms comply with environmental and animal welfare regulations. ◯

Organic Farming

5 a) What is organic farming? ....

....

    b) What percentage of UK farmland is now organic? .... ▢

Environmental Impact of Modern Farming

6 The diagram shows nitrate pollution from artificial fertilizers. Match statements **A, B, C, D** and **E** with the labels **1–5** on the diagram. Enter the appropriate number in the boxes provided.

A Nitrates washed over ground surface. ⬭   **B** Aquatic life dies. ⬭

C Algae grows and uses up oxygen. ⬭   **D** Nitrates sprayed on fields. ⬭

E Nitrates collect in rivers and lakes. ⬭

7 Choose the correct words from the options given to complete the following sentences.

habitats	decrease	hedgerows	machinery

Over the last 60 years 400 000km of _____ have been removed to make fields

larger for _____. Important _____ for wildlife are disappearing

and there has been a _____ in insects, birds and animals.

Government Policies to Reduce Environmental Impacts

8 The table contains three terms relating to policies that the UK government has introduced to reduce the environmental impact of agriculture. Match descriptions **A, B** and **C** with the terms **1–3** in the table. Enter the appropriate number in the boxes provided.

1	Single Payment Scheme
2	Environmental Stewardship Scheme
3	Regulations

A Farmers are paid to conserve wildlife, protect the environment and improve the landscape. ⬭

B The Department for Environment, Food and Rural Affairs controls the environmental impact of farming by ensuring that farmers follow environmental rules and laws. ⬭

C Farmers are paid money to keep their land in good environmental condition. ⬭

Farming in the Tropics

Traditional Farming and Loss of Rainforest

1 Choose the correct words from the options given to complete the following sentences.

| subsistence | ash | 50 | fertility | 4–5 | cultivation |

a) _____ farming is when crops are grown, or animals reared, to provide food for the producer's family.

b) Shifting _____ is when farmers move from place to place every _____ years.

c) _____ is used to fertilise the soil.

d) The farmers move when the soil loses its _____ .

e) It takes about _____ years for the land to recover and the vegetation to grow back.

2 The diagram shows the four stages of shifting cultivation. Match statements **A**, **B**, **C** and **D** with the labels **1–4** on the diagram. Enter the appropriate number in the boxes provided.

A Crops are cultivated. ( )

B Tropical rainforest. ( )

C Forest regenerates. ( )

D Vegetation is cleared. ( )

Cash Crops

3 Which of the following statements about cash crops are true? Tick the correct options.

A A cash crop is grown to make money. ( )

B Most cash crops are grown on large farms known as plantations. ( )

C Common cash crops are yams, cassava and sweet potatoes. ( )

D Transnational companies often own farms that grow cash crops. ( )

E A monoculture is when more than one type of crop is grown on a farm. ( )

4 Describe two disadvantages of plantation agriculture.

a) _____

( )

b) _____

Farming in the Tropics

Agricultural Change – Green Revolution

5 Which of the following describes the Green Revolution? Tick the correct option.

A An increase in awareness and concern for the environment. ◯

B Population growth in poorer tropical countries during the 1960s and 1970s. ◯

C A strategy aimed at increasing crop yields by introducing new technology. ◯

D A political movement led by farmers, which aimed to take over the governments of some poorer tropical countries. ◯

6 Fill in the missing words to complete the following sentences.

During the Green Revolution, _____ yield varieties of wheat and rice were developed

that grew faster and produced a bigger crop. Farmers had to buy _____ to replace

nutrients lost from the soil, and _____ to enable them to farm a larger area more

efficiently. Yields increased by _____ and some farmers found they could produce

_____ crops a year. However, poorer farmers couldn't _____ new

crops or technology, and the chemicals used caused _____ .

Agricultural Change – Irrigation

7 a) What is irrigation?

....

b) Why is irrigation necessary in sub-tropical areas?

....

Agricultural Change – Appropriate Technology

8 Draw lines between the boxes to match each type of appropriate technology with its description.

Intercropping		Leguminous crops that add nitrogen to the soil
Food storage		Growing a variety of plants to maximise productivity
Natural fertilizer		Prevention of crop loss by rodents, insects or disease

Contrasts in Development

1 The table contains eight indicators of development.

Match descriptions **A, B, C, D, E, F, G** and **H** with the terms **1–8** in the table. Enter the appropriate number in the boxes provided.

1	Gross National Product (GNP)
2	Gross National Income per head (GNI)
3	Birth rate
4	Death rate
5	Infant mortality rate
6	People per doctor
7	Access to safe water
8	Life expectancy

A The percentage of people who have a clean water supply.

B The number of babies born per thousand people per year.

C The total value of goods and services produced by a country in a year.

D The average life span in years.

E The number of deaths per thousand people per year.

F The number of babies who die before they're one year old per thousand live births.

G The population divided by the number of doctors.

H The total value of goods and services produced by a country in a year, divided by its population.

2 Which of the following statements about the Human Development Index (HDI) are true? Tick the correct options.

A The HDI is calculated using measures of health, education and standard of living.

B The HDI of a country is indicated by a score from 0 to 1.

C The countries with the highest HDI tend to be in Africa.

D As a country becomes more developed its HDI score becomes closer to 1.

Contrasts in Development

3 The diagram shows the global North–South divide. Match statements **A, B, C, D, E, F, G** and **H** with the labels **1–8** on the diagram. Enter the appropriate number in the boxes provided.

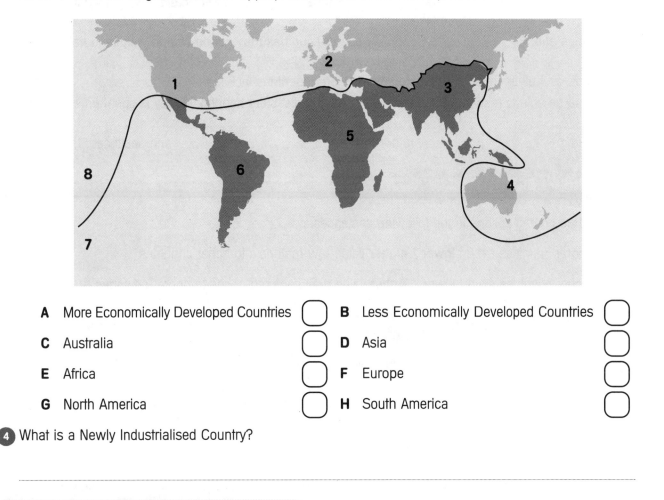

A   More Economically Developed Countries ⬭   B   Less Economically Developed Countries ⬭

C   Australia ⬭   D   Asia ⬭

E   Africa ⬭   F   Europe ⬭

G   North America ⬭   H   South America ⬭

4 What is a Newly Industrialised Country?

_____

Quality of Life

5 Choose the correct words from the options given to complete the following sentences.

perception   well-being   services   health   goods   happiness

a) Quality of life is a measure of _____ that includes the _____

and _____ of people.

b) Standard of living is a measure of the quality of _____ and

_____ available to people.

c) Quality of life is based on people's _____, so the definition of an
acceptable quality of life may vary from one country to another.

Global Inequalities

Environmental Factors

1 Choose the correct words from the options given to complete the following sentences.

rebuilding   **resources**   **education**   **infrastructure**   **healthcare**

Buildings and .... in poorer countries may not be able to withstand natural

hazards as well as those in richer countries. Poorer countries lack the .... and
funds to be able to respond adequately when natural disasters happen. Money is spent on emergency

aid and .... rather than on .... and healthcare.

Economic Factors

2 Circle the correct options in the following sentences.

a) Poorer countries export **lower / higher** value raw materials to richer countries.

b) Richer countries export **lower / higher** value manufactured goods to poorer countries.

Social Factors

3 Fill in the missing words to complete the following sentences.

In poorer countries .... billion people do not have access to a safe water supply.

Untreated water may contain diseases, which can lead to sickness or even ....

If people are unwell, they can't work and they put a strain on .... services.

Political Influences

4 The table contains three development obstacles associated with some poorer countries. Match descriptions
A, **B** and **C** with the terms **1–3** in the table. Enter the appropriate number in the boxes provided.

1	War
2	Corruption
3	Incompetence

A Governments may lack the knowledge or experience to run a country effectively.

B Civil and international conflict costs governments money.

C Government ministers taking bribes to give contracts to particular companies.

Inequalities in the European Union

5 a) What is the European Union?

_____

b) Name three countries in the European Union with a high Human Development Index.

i) _____ **ii)** _____ **iii)** _____ .

c) Name three countries in the European Union with a low Human Development Index.

i) _____ **ii)** _____ **iii)** _____ .

6 Which of the following statements about the more developed nations of the European Union are true? Tick the correct options.

A They are found towards the East of the European Union.

B They are found at the core of Europe.

C They have democratically elected governments.

D They have a history of being controlled by communist governments.

E They have modern industries.

F They have traditional industries.

G They have economies that are based around tertiary activities.

European Union Policies

7 How big is the European Union budget? Tick the correct option.

A $100 million

B $120 billion

C $200 million

D $210 billion

8 Fill in the missing words to complete the following sentences.

a) The _____ Policy is designed to give farmers a fair standard of living by controlling prices for produce and encouraging sustainable farming practices.

b) _____ is the movement of money from the richer countries of the European Union to the poorer countries in order to stimulate economic growth and development.

Reducing Global Inequalities

World Trade

1 Which of the following statements about world trade are true? Tick the correct options.

A International trade is the exchange of goods and services between countries. ◯

B The pattern of international trade is currently very unequal. ◯

C Trade happens when a producing country is able to produce goods of a higher quality or lower price than a consuming country. ◯

D A quarter of global trade happens between the eight richest countries in the world, known as the G8. ◯

E China and India have yet to play a major role in world trade. ◯

Trading Groups

2 Fill in the missing words to complete the following sentences.

Within a trading group, the countries will reduce import taxes known as ....

They'll also remove .... , which are limits placed on the imports that a country usually allows. The trading group will then make it harder for countries outside the

group to .... their goods to the members of the trading group.

3 The table contains the names of five trading groups. Match descriptions **A, B, C, D** and **E** with the world trading groups **1–5** in the table. Enter the appropriate number in the boxes provided.

1	European Union
2	North American Free Trade Agreement
3	Mercosur
4	African Free Trade Association
5	Association of South East Asian Nations

A The USA, Canada and Mexico make up this trade group. ◯

B This trade group has 10 member countries including Singapore and Thailand. ◯

C There are 26 member countries of this trade group including Kenya and Tanzania. ◯

D This trade group is formed from countries in South America. ◯

E Belgium, France, Italy, Luxembourg, the Netherlands and Germany were the first members of this trade group. ◯

Reducing Global Inequalities

Fair Trade

4 What is meant by the term 'fair trade'?

_____

_____

5 Write down two benefits of fair trade to producers in poorer countries.

a) _____

b) _____

International Aid

6 Circle the correct options in the following sentences.

a) Aid is the transfer of resources from a **richer / poorer** country to a **richer / poorer** country.

b) Aid that is from an agreement between two countries is known as **bilateral / multi-lateral** aid.

c) The International Monetary Fund, United Nations and World Bank are regular donors of **bilateral / multi-lateral** aid.

d) **Short-term / Long-term** aid is help given to countries in the immediate aftermath of a natural disaster.

e) Aid that is intended to improve the standard of living of a country is known as **short-term / long-term** aid.

f) **Debt abolition / Conservation swap** is debt that is paid for by environmental groups on the condition that a country undertakes projects that protect nature.

Case Study: Asian Tsunami, 2004

7 Which of the following responses to the 2004 Asian Tsunami by Oxfam are examples of short-term aid? Tick the correct options.

A Wells were cleaned and repaired. ◯

B Emergency aid was flown in by helicopter. ◯

C Radio was used to keep local people up-to-date with the aid that was being provided. ◯

D Houses have been re-built. ◯

E Fishermen have been provided with new fibre-glass fishing boats. ◯

F Local people were paid to clear mud from streets and homes. ◯

Globalisation and Industry

Globalisation and Interdependence

1 Fill in the missing words to complete the following sentences.

Globalisation is the increase in _____, social and political links between countries

since the _____. Many countries are now _____, which means
that they rely on each other as trading partners.

Information Communication Technology

2 Which of the following statements about Information Communication Technology are true? Tick the correct options.

A Computers are becoming cheaper and more powerful. ◯

B Most people prefer to use more traditional means of communication, such as a letter. ◯

C The cost of data transmission has fallen due to the creation of a global fibre-optic cable network. ◯

D Many large companies have set up telephone and internet based customer service centres, known as call centres, in countries like India and the Philippines. ◯

Transnational Corporations

3 a) What is a transnational corporation?

_____

b) Where are the headquarters of transnational corporations located?

_____

c) Give three advantages of transnational corporations.

i) _____

ii) _____

iii) _____

d) Give three disadvantages of transnational corporations.

i) _____

ii) _____

iii) _____

Globalisation and Industry

4 The table contains five terms relating to the company, Nike. Match descriptions **A, B, C, D** and **E** with the processes **1–5** in the table. Enter the appropriate number in the boxes provided.

1	Research, design and marketing
2	Manufacturing
3	Subcontracting
4	Exploitation
5	Profits

A In 2007 Nike made $1.5 billion. ◯

B This is located in the USA. ◯

C Allegations of treating workers badly by paying low wages for work in poor conditions. ◯

D A system of paying another company to provide goods and services for a company. ◯

E This is located in other countries, mostly poorer than the USA. ◯

Changes in Manufacturing

5 What is deindustrialisation?

....

....

6 Circle the correct options in the following sentences.

a) Newly industrialised countries are able to compete with richer, more developed, countries because their wages are **lower / higher**, so they can sell their manufactured goods at a **cheaper / more expensive** price.

b) Raw materials are **cheaper / more expensive** in poorer countries.

c) Globalisation has made it easier for countries to buy from the **cheapest / closest** producer.

d) People are prepared to work for **long hours / low pay** in newly industrialised countries.

e) Unions and strikes are **illegal / encouraged**.

f) The governments of newly industrialised countries are prepared to give tax incentives to encourage **people / transnational corporations** to move to their country.

Globalisation and Energy

Increasing Demand for Energy

1 World energy use is expected to have doubled by which year? Tick the correct option.

A 2030 ◯

B 2050 ◯

C 2070 ◯

D 2080 ◯

2 **a)** What is the current approximate population of the world? ....

b) What is the world's population expected to reach by 2050? ....

3 Which of the following statements about world energy use are true? Tick the correct options.

A The population of India and China combined is nearly 1.5 billion people. ◯

B As countries develop, their demand for goods, and therefore energy, increases. ◯

C There's a huge demand for high-tech goods that use a lot of energy. ◯

D Appliances are being designed with little regard for their energy use, so the demand for energy is going to continue to increase. ◯

E Many items have now been developed to be more energy-efficient. ◯

Impacts of Increased Energy Use

4 Fill in the missing words to complete the following sentences.

a) Land, air and water is at risk of .... from burning ....

fuels or radiation leaks from .... power. But there is ....
over the siting of renewable energy projects.

b) The cost of fuel and electricity may .... as non-renewable resources begin to
run out.

c) People on .... incomes could suffer as they may not be able to afford high

energy ....

Sustainable Energy Use

5 (Circle) the correct options in the following sentences.

a) Sustainable development means meeting the needs of the **present / future** generation without limiting the ability of **present / future** generations to meet their own needs.

b) The Kyoto Protocol is an agreement signed at the Earth Summit held in **China / Japan** in **1997 / 2007.**

Renewable Energy

6 The table contains the names of five types of renewable energy. Match descriptions **A, B, C, D** and **E** with the energy types **1–5** in the table. Enter the appropriate number in the boxes provided.

1	Hydroelectric power
2	Wind energy
3	Solar energy
4	Geothermal energy
5	Tidal energy

A Electricity that's generated using steam, produced by boiling water that uses heat from the Earth's mantle.

B Electricity generated from turbines turned by the movement of air.

C Electricity generated by harnessing the energy produced by the pull of the moon on coastal waters.

D Electricity generated by flowing water being channelled through turbines.

E Energy from the sun that's used to heat water or generate electricity.

Local to Global Action

7 Draw lines between the boxes to match the scale with the type of action.

| Local |

| National |

| Global |

| Governments can introduce policies to encourage people to save energy. |

| Governments can sell their quota of greenhouse gas emissions to other countries. |

| Putting waste in a recycling bin is an effective way for everyone to save energy. |

Globalisation and Food

Demand for Food

1 Circle the correct options in the following sentences.

a) The global demand for food is expected to double by **2025 / 2050**.

b) The world's population is growing by about **75 / 125** million people a year.

c) Increasing wealth in richer countries has led to an increase in demand for **fresh / processed** foods.

d) Large-scale farming that produces food as cost effectively as possible is known as **industrialised / sustainable** farming.

Environmental Impacts

2 Which of the following statements about importing food are true? Tick the correct options.

A Poorer countries import more food than they export to richer countries. ◯

B Half of the vegetables sold in the UK are imported. ◯

C 95% of the fruit sold in the UK is imported. ◯

D Consumers are demanding more exotic foreign foods. ◯

E Consumers want to be able to buy fruit and vegetables all year round. ◯

F The National Farmers' Union is working hard to encourage more people to buy food from abroad. ◯

G Farmers in poorer countries often have to farm on low quality marginal land because better farmland is used to grow crops to export to richer countries. ◯

H Soil erosion and salinisation are both problems caused by over-farming marginal land. ◯

3 a) Explain what is meant by the term 'food miles'.

_____

_____

b) Explain what is meant by the term 'carbon footprint'.

_____

_____

Globalisation and Food

Social Impacts

4 **a)** What is subsistence farming?

....

b) Why have many farmers changed from subsistence farming to growing cash crops?

....

c) How does fair trade help poor farmers in developing countries?

....

d) How many farmers in poor countries are poisoned by pesticides every year?

....

Political Impacts

5 Fill in the missing words to complete the following sentences.

a) .... is a system of taking water from one place and using it to water crops in another place.

b) There can be .... between groups over access to water supplies. In some

places .... is likely to run out as people are taking water out faster than it's being filled up naturally.

c) By .... the majority of the world's population will live in an area with a water shortage.

Economic Impacts

6 Choose the correct words from the options given to complete the following sentences.

supermarkets	**costs**	**200**	**taxes**

The UK spends £.. million a year on fruit and vegetables from Africa, and the

money goes to benefit local people. .... raised by cash crops can be used to improve healthcare, education and infrastructure. Chemicals and fertilizers can greatly increase a

farmer's .... Most profits are made by the .... , such as Tesco and Sainsbury's.

Growth of Tourism

Global Increase in Tourism

1 Fill in the missing words to complete the following sentences.

Tourism is when people visit places for _____. It may range from a single

_____ away, to a holiday of several _____.

2 Which of the following statements about the growth of tourism are true? Tick the correct options.

- **A** There are approximately 700 million international tourist visits every year in the world. ◯
- **B** The number of international tourist visits is expected to rise to over 1.5 billion by 2020. ◯
- **C** Employers are giving workers less time off for holidays. ◯
- **D** People are being paid more, so they can afford to go on more holidays. ◯
- **E** Countries with ageing populations have fewer tourists because older people tend to stay at home more than younger people. ◯
- **F** Travel programmes on television encourage people to go on holiday. ◯
- **G** Air travel is becoming increasingly expensive. ◯

Tourism Potential

3 The table contains the names of three types of tourist destination. Match descriptions **A, B, C, D, E** and **F** with the types of holiday **1–3** in the table. Enter the appropriate number in the boxes provided (use each number twice).

1	Cities
2	Mountains
3	Coasts

- **A** Tourists come here for walking, climbing and skiing. ◯
- **B** Paris, New York, Dubai, Rome and Prague. ◯
- **C** Barbados, Phuket, Ibiza and Cornwall. ◯
- **D** Tourists come here for sightseeing, entertainment and shopping. ◯
- **E** The Alps, the Rockies and the Himalayas. ◯
- **F** Tourists come here for sunbathing, watersports and the nightlife. ◯

Economic Importance

4 Circle the correct options in the following sentences.

a) Tourism is the world's largest industry, with a turnover of approximately $700 **million / billion**.

b) The majority of tourist visits take place in **richer / poorer** countries.

c) The country which receives the most money from tourism is **France / the USA**.

Life Cycle Model

5 The diagram shows the resort life cycle model. Match statements **A, B, C, D, E** and **F** with the labels **1–6** on the diagram. Enter the appropriate number in the boxes provided.

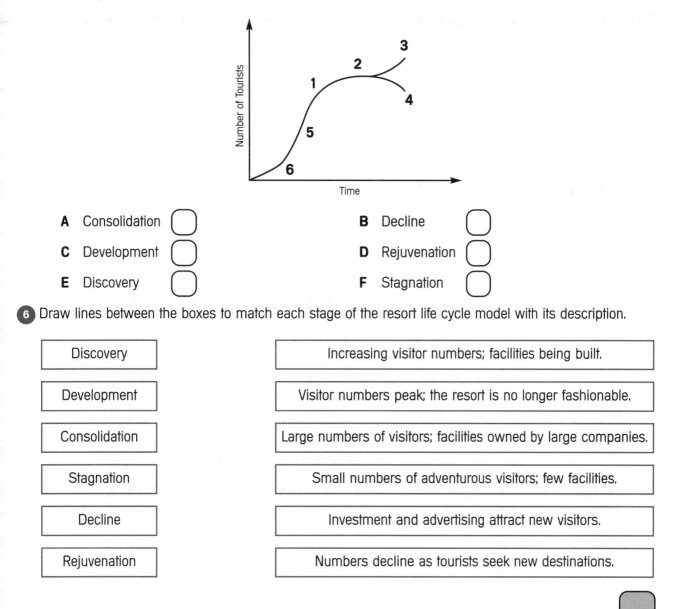

A Consolidation	⬜	**B** Decline	⬜
C Development	⬜	**D** Rejuvenation	⬜
E Discovery	⬜	**F** Stagnation	⬜

6 Draw lines between the boxes to match each stage of the resort life cycle model with its description.

Discovery	Increasing visitor numbers; facilities being built.
Development	Visitor numbers peak; the resort is no longer fashionable.
Consolidation	Large numbers of visitors; facilities owned by large companies.
Stagnation	Small numbers of adventurous visitors; few facilities.
Decline	Investment and advertising attract new visitors.
Rejuvenation	Numbers decline as tourists seek new destinations.

Tourism in the UK

Tourism and the UK Economy

1 Choose the correct numbers from the options given to complete the following sentences.

2.1	15	30	5

a) Tourism is the UK's _____th largest industry.

b) Tourism employs _____ million people in the UK.

c) Over _____ million people visit the UK every year, contributing

£_____ billion to the economy.

2 Give three reasons why the number of tourist visits to the UK fell at the start of the 21st Century.

a) _____

b) _____

c) _____

Case Study: The Lake District

3 Fill in the missing words to complete the following sentences.

a) The Lake District is a _____ _____ in the North West of England.

Approximately _____ million people visit the Lake District every year.

b) Visitors are attracted by the spectacular scenery of _____ and

_____, outdoor activities such as _____ and

_____, picturesque towns including _____ and

_____, and many purpose-built tourist attractions.

4 Write down three ways in which the Lake District management plan is aiming to make tourism in the Lake District more successful.

a) _____

b) _____

c) _____

Tourism in the UK

Case Study: The Lake District (cont.)

5 Which of the following statements about tourism in the Lake District are true? Tick the correct options.

A Lake Windermere and Keswick have been allowed to develop as key tourist destinations, so that other parts of the Lake District receive fewer tourists. ◯

B A 'honeypot site' is a place where tourists aren't allowed to visit so that the environment or heritage is protected. ◯

C Footpaths are being eroded by walkers and heavy rain. ◯

D Too many cars and tourist coaches cause congestion and pollution. ◯

E Traffic is being controlled by a congestion charge, which visitors are required to pay before they enter the region. ◯

Case Study: Pembrokeshire

6 Circle the correct words from the options given to complete the following sentences.

a) Tenby is a **coastal / mountain** resort in South West Wales.

b) Tenby has 4km of **sandy beaches / steam railway lines**.

c) A major tourist attraction to Tenby is its **theme park / medieval walls**.

d) Shops, pubs and restaurants are geared to **tourists / local people**.

7 Which of the following statements about tourism in Tenby are true? Tick the correct options.

A The town centre has been pedestrianised. ◯

B A park-and-ride scheme has been set up to reduce congestion. ◯

C A new shopping complex with luxury hotels has been built in the centre of Tenby. ◯

D Recycling facilities have been put on the beach. ◯

E Tenby is marketed as a summer destination to reduce the pressure of tourism in the winter. ◯

F Tourist information centres have been located in key destinations in Pembrokeshire, including Tenby. ◯

G The Celtic Trail, a long-distance cycle path, has been set up. ◯

Mass Tourism and Ecotourism

Mass Tourism

1 Circle the correct options in the following sentences.

   a) Mass tourism expanded rapidly in the **1960s / 1970s**.

   b) A major reason for the rapid expansion of mass tourism was the fall in the price of **air travel / petrol**.

   c) Holidays where travel, accommodation and, sometimes, food are paid for in advance are known as **excursions / package holidays**.

Case Study: Kenya

2 Fill in the missing words to complete the following sentences.

Kenya is a .... country on the .... coast of Africa. Mass

tourism developed in Kenya in the .... Tourists are attracted to Kenya because

it has beautiful ...., such as the savannah grasslands and coral beaches. Tourists

also come to Kenya to visit at least one of the 19 ....

3 Which of the following statements about tourism in Kenya are true? Tick the correct options.

   A Kenya receives 1 million tourists a year. ◯

   B Tourism accounts for 50% of Kenya's income. ◯

   C 500 000 people earn an income directly from tourism. ◯

   D The Kenya Wildlife Service gets its money from the government. ◯

   E Local people don't benefit from tourism as much as they could because often
   profits from tourism go to international hotel chains and holiday companies. ◯

4 Choose the correct words from the options given to complete the following sentences.

palm trees	Lamu	buildings	future	current	local

   a) Sustainable tourism is all about meeting the needs of .... tourists, without

   affecting the ability of .... tourists to meet their own needs.

   b) In .... tourists are encouraged to stay in guest houses run by

   .... people, new buildings must be no higher than ....

   and a tourist tax is paid to conserve historic ....

Mass Tourism and Ecotourism

5 Which of the following describes ecotourism? Tick the correct option.

A Tourism that involves camping, trekking and climbing mountains. ◯

B Tourism to natural areas, which conserves the environment and improves the well-being of local people. ◯

C Tourism that involves staying in a hotel where everything is paid for in advance. ◯

D Tourism where the tourists save money. ◯

Case Study: Antarctica

6 a) When did tourism to Antarctica begin? ....

b) How many tourists visit Antarctica each year? ....

c) What time of year do most tourists go to Antarctica? ....

d) Why do tourists go to Antarctica?

....

e) Give three positive impacts of tourism in Antarctica.

i) ....

ii) ....

iii) ....

f) Give three negative impacts of tourism in Antarctica.

i) ....

ii) ....

iii) ....

7 ⟨Circle⟩ the correct options in the following sentences.

a) Tourists are only allowed to come ashore in groups of **10 / 100**.

b) Tourists mustn't take any **photographs / souvenirs**.

c) Tourist ships aren't allowed to **discharge waste / put down their anchors**.

Exam-style Questions

1 The Restless Earth

1 (a) (i) Explain why volcanoes are found at constructive plate boundaries.

....

....

.... *(3 marks)*

1 (a) (ii) Describe the ways in which a shield volcano is different from a composite volcano.

....

....

....

.... *(4 marks)*

1 (b) Study **Figure 1**, the map that shows the location of the epicentre of the earthquake that caused the 2004 Asian tsunami.

Figure 1

1 (b) (i) Describe the causes and effects of a tsunami that you have studied.

....

....

....

....

.... *(8 marks)*

2 Challenge of Weather and Climate

2 (a) Study **Figure 2**, a map showing the average annual rainfall in the UK.

Figure 2

- over 2000mm
- 750-2000mm
- under 750mm

Fort William 2000mm ·

Glasgow 1560mm ·

Newcastle ·630mm

Keswick 1480mm ·

Manchester 860mm ·

London 610mm

Falmouth 1100mm

2 (a) (i) Describe the pattern of the average annual rainfall in the UK.

....

.... *(3 marks)*

2 (a) (ii) Explain why Keswick has more rainfall than Newcastle.

....

....

.... *(4 marks)*

2 (b) (i) Study **Figure 3**, on page 87, the photograph of cyclone damage in Bangladesh.

Describe the effects on people of tropical storms in poor countries like Bangladesh.

....

.... *(3 marks)*

2 (c) (i) Using case studies of tropical storms in rich and poor parts of the world, describe the contrasting immediate and long-term responses to the storms.

....

....

....

....

.... *(6 marks)*

Exam-style Questions

3 Living World

3 (a) Study **Figure 4**, the map showing the global distribution of ecosystems.

Figure 4

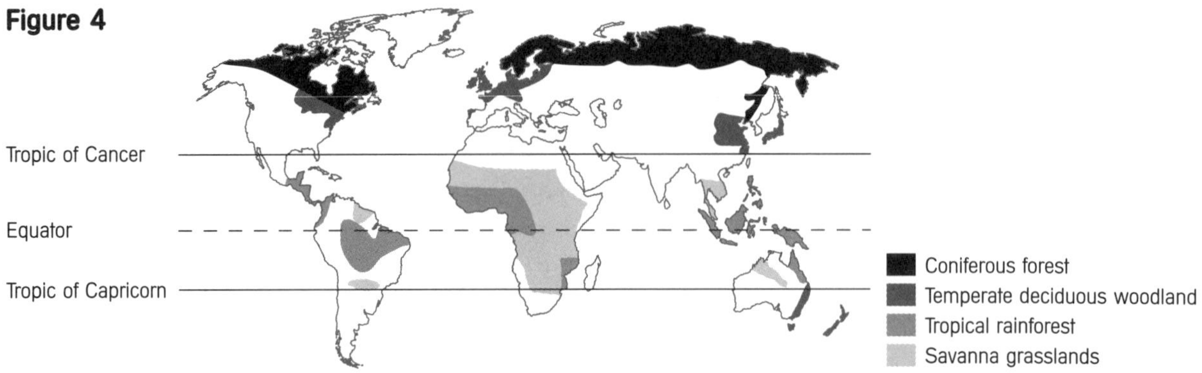

Tropic of Cancer

Equator

Tropic of Capricorn

- ■ Coniferous forest
- ■ Temperate deciduous woodland
- ■ Tropical rainforest
- ■ Savanna grasslands

3 (a) (i) Describe the location of the world's deserts.

....

....

.... *(3 marks)*

3 (a) (ii) Explain how the vegetation of hot deserts has adapted to its environment.

....

....

.... *(4 marks)*

3 (b) Study **Figure 5**, on page 87, the photograph of Las Vegas, which is located in the Mojave Desert, in the South-western USA.

3 (b) (i) Describe the advantages of hot deserts for tourism.

....

....

.... *(4 marks)*

3 (b) (ii) Describe the disadvantages of hot deserts for tourism.

....

....

.... *(4 marks)*

4 Water on the Land

4 (a) Study **Figure 6**, a diagram that shows the lower course of a river.

Figure 6

B _____

C _____

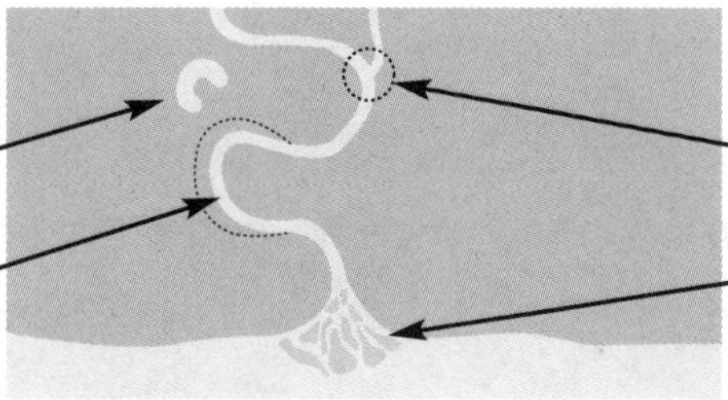

A _____

D _____

4 (a) (i) Label features **A–D** on the diagram. *(4 marks)*

4 (a) (ii) Describe and explain the formation of levees.

_____

_____

_____

_____

_____

_____ *(6 marks)*

4 (b) (i) Describe the human and physical causes of a flood you have studied.

_____

_____

_____ *(4 marks)*

4 (c) (i) Describe the advantages of hard engineering as a means of controlling flooding.

_____

_____

_____ *(4 marks)*

4 (c) (ii) Describe the disadvantages of hard engineering as a means of controlling flooding.

_____

_____

_____ *(4 marks)*

Exam-style Questions

5 The Coastal Zone

5 (a) Study **Figure 7**, a diagram showing a destructive wave.

Figure 7

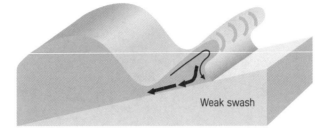

Weak swash

5 (a) (i) Contrast the characteristics of destructive and constructive waves.

....

....

.... *(3 marks)*

5 (a) (ii) Explain the formation of a wave-cut platform.

....

....

....

.... *(6 marks)*

5 (b) (i) For a case study of coastal management you have studied, assess the costs and benefits of the strategies adopted.

....

....

....

....

....

.... *(8 marks)*

Human Geography

1   Changing Urban Environments

1   (a) Squatter settlements are areas of self-built housing usually found on the edge of urban areas in poorer countries on land that does not belong to the people who live there.

1   (a) (i)   Describe the living conditions of people who live in squatter settlements.

....

....

.... *(3 marks)*

1   (a) (ii) Using a named example, describe how the government has tried to improve the lives of people living in a squatter settlement.

....

....

....

....

....

.... *(6 marks)*

1   (a) (iii) Using a named settlement you have studied, explain how planners are attempting to make the settlement more sustainable.

....

....

....

.... *(4 marks)*

Exam-style Questions

2 **Changing Rural Environments**

2 **(a)** Study the Ordnance Survey map of Witney on page 88. A major supermarket is considering building a £40 million superstore on the rural–urban fringe of Witney. The site they have selected is in grid square 3309.

Using evidence from the Ordnance Survey map, give two reasons why this is a suitable site for a superstore.

2 **(a) (i)** ....

....

....

....

.... *(4 marks)*

2 **(b) (i)** Name one group of people who might object to a superstore being built there.

.... *(1 mark)*

2 **(b) (ii)** Why is this group of people against developments on the rural–urban fringe?

....

....

....

.... *(3 marks)*

2 **(c)** Oxfordshire County Council is proposing to build a £15 million link road in grid square 3609, from the roundabout at 359092 to Cogges.

Use evidence from the Ordnance Survey map of Witney on page 88 to answer the following questions.

2 **(c) (i)** Suggest one group of people who might support the proposed link road.

.... *(1 mark)*

2 **(c) (ii)** Give two reasons why the land in grid square 3609 has not been previously developed.

....

....

.... *(2 marks)*

3   Globalisation

3   (a) Read **Figure 8**, the newspaper article.

Figure 8

> # UKPhone to open call centres in India
>
> UKPhone, like many other British companies, is planning to open call centres in India. 2 000 jobs in the UK could be lost.
>
> UKPhone would like to open a call centre in Bangalore, and another in Delhi. UKPhone employs 16 000 people in 100 call centres in the UK.
>
> This number will fall to 14 000 people in just 31 call centres if the plans to relocate go ahead.
>
> UKPhone claims that no UK employees will be made redundant and no agency staff will be laid off. However, unions are up in arms over yet another transfer of UK jobs overseas.

3   (a) (i)   What is meant by the term 'call centre'?

.... *(1 mark)*

3   (a) (ii)   Explain why companies like BT are choosing to locate their call centres in countries like India rather than the UK.

....

....

.... *(4 marks)*

3   (b) (i)   Describe the advantages and disadvantages of TNCs.

....

....

....

.... *(6 marks)*

3   (c)   The global demand for energy is expected to double by 2030.

3   (c) (i)   Account for the rise in the global demand for energy.

....

....

.... *(3 marks)*

Exam-style Questions

4   **Tourism**

4   **(a)** Study **Figure 9**, on page 87, a photograph of the River Dove, a honeypot site in the Peak District National Park.

4   **(a) (i)** Use the photograph to suggest why this area is so popular with tourists.

....

.... *(2 marks)*

4   **(a) (ii)** Describe the problems that tourism can bring for people and the environment for an area such as the River Dove.

....

....

....

.... *(4 marks)*

4   **(b) (i)** For a National Park that you have studied, describe the strategies that have been put in place to ensure that tourism continues to be a success in the area.

....

....

....

....

.... *(6 marks)*

4   **(c)** Choose one extreme environment that attracts tourists.

4   **(c) (i)** Describe the advantages that tourism brings to this area.

....

.... *(4 marks)*

4   **(c) (ii)** Describe the disadvantages that tourism brings to this area.

....

.... *(4 marks)*

Images for use with Exam-style Questions

Figure 3: Photograph of cyclone damage in Bangladesh

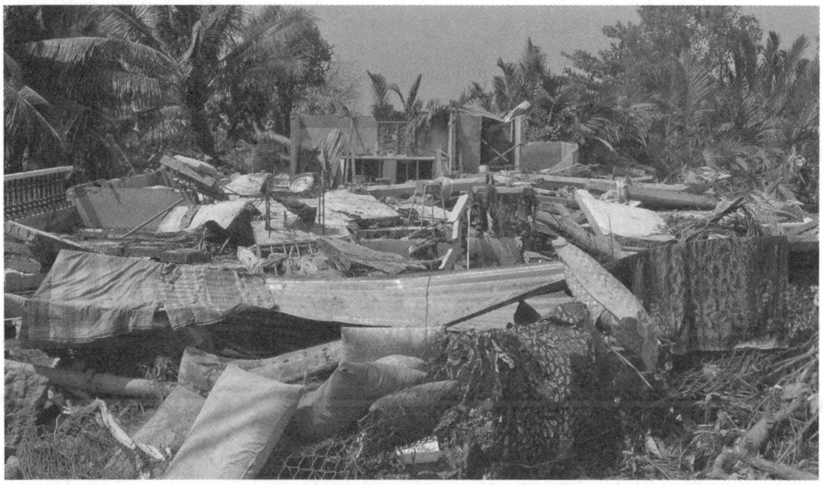

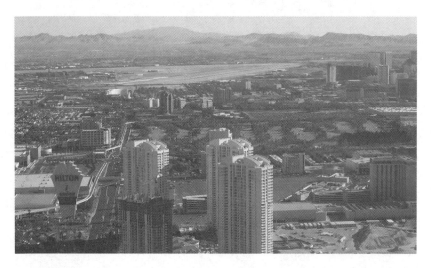

Figure 5: Photograph of Las Vegas, located in the Mojave Desert, South-western USA

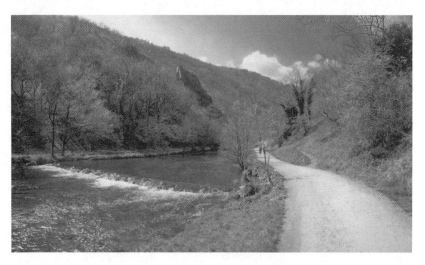

Figure 9: Photograph of the River Dove, a honeypot site in the Peak District National Park

Ordnance Survey Map

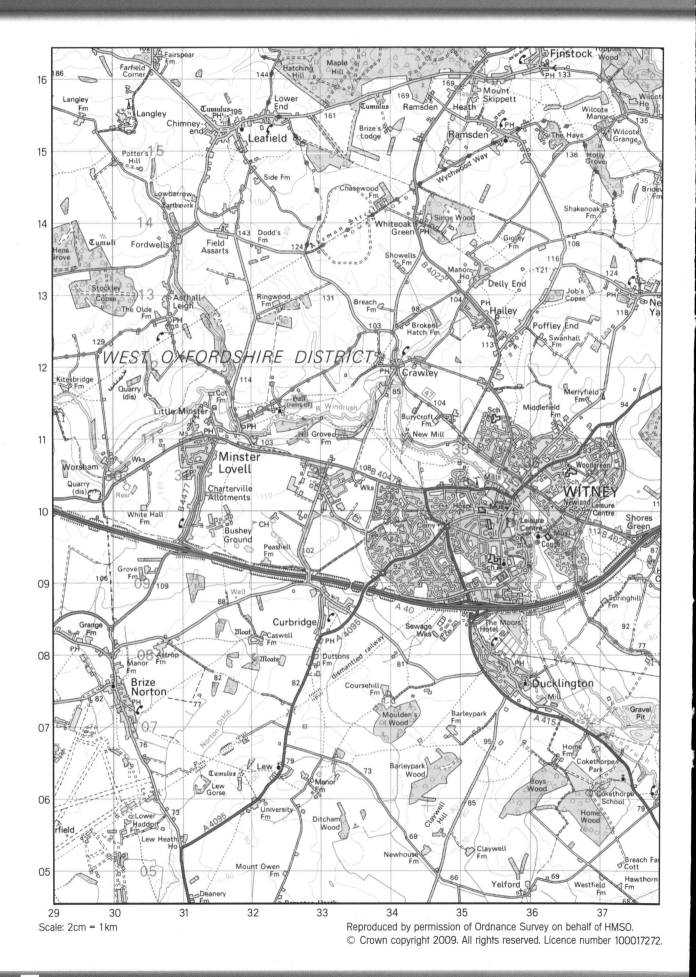

Scale: 2cm = 1km